mange tout™

mange tout™

Teaching your children to love fruit and vegetables without tears

LUCY THOMAS

PENGUIN

MICHAEL
JOSEPH

For Josh, Chloe and Will,
Perfect as Three Peas in a Pod
With Lots of Love, Laughter and Vegetable Sauce!

MICHAEL JOSEPH

Published by the Penguin Group
Penguin Books Ltd, 80 Strand, London WC2R 0RL, England
Penguin Group (USA) Inc., 375 Hudson Street, New York, New York 10014, USA
Penguin Group (Canada), 90 Eglinton Avenue East, Suite 700,
Toronto, Ontario, Canada M4P 2Y3 (a division of Pearson Penguin Canada Inc.)
Penguin Ireland, 25 St Stephen's Green, Dublin 2, Ireland (a division of Penguin Books Ltd)
Penguin Group (Australia), 250 Camberwell Road,
Camberwell, Victoria 3124, Australia (a division of Pearson Australia Group Pty Ltd)
Penguin Books India Pvt Ltd, 11 Community Centre,
Panchsheel Park, New Delhi – 110 017, India
Penguin Group (NZ), 67 Apollo Drive, Rosedale, North Shore 0632, New Zealand
(a division of Pearson New Zealand Ltd)
Penguin Books (South Africa) (Pty) Ltd, 24 Sturdee Avenue,
Rosebank, Johannesburg 2196, South Africa

Penguin Books Ltd, Registered Offices: 80 Strand, London WC2R 0RL, England

www.penguin.com

First published 2007
1

Text copyright © Lucy Thomas, 2007
Illustrations copyright © Ana Maria, 2007

The moral right of the author has been asserted

Designed by seagulls.net

Printed in Italy by Printer Trento, s.r.l.

A CIP catalogue record for this book is available from the British Library

ISBN: 978-0-718-15294-9

Contents

About the Author

Mange Tout was born out of my passion for children, performance, health and nutrition. I trained as an actress and have had ten years experience as a professional nanny. A lifelong interest in health and nutrition also led to a diploma in Food and Nutrition at the School of Natural Health Sciences in London.

In my role as a nanny, I noticed that the more opportunities children were given to be involved in activities such as shopping, food handling and preparation and even simple gardening, the more relaxed and interested they became in fresh produce. I also realized that children are less inhibited and far more likely to sample a new food if they are not asked to eat or taste it. By first showing a child and then asking them if they can tickle their tongue with a piece of broccoli, they will more often than not get excited and respond with a positive action. So they are not only handling what might be a new food, but also introducing themselves to a new taste, even if on a very simple level.

I started out by writing easy songs and rhymes about fruit and vegetables with simple actions to accompany them. We all know how much children love making a noise and being active, so by giving them the chance to let off some steam by singing the 'Hokey Cokey Celery' they were then geared up and excited to learn about fruit and vegetables. Today, through my Mange Tout classes and now this book, I hope to give many more parents and children the exciting opportunity to develop healthy eating habits and have fun as well.

Lucy Thomas

Foreword

One dull May morning in 2006, I visited a Mange Tout class in south London. As soon as I stepped through the door and saw a group of preschoolers running round the room searching for fruit and vegetables, a light went on. I realized this was exactly what was needed for so many of the children I had been treating at Great Ormond Street Hospital. I also had a sneaking suspicion that I would go home and try out her ideas on my little ones, both of whom had just reached that fussy stage.

I had been working with children with feeding difficulties for some years as a clinical psychologist, and had long wished for a good preventative programme to manage small children's selective eating habits. Here it was! I knew that if this class had been locally available, for many children a long wait to see a paediatrician, dietician or psychologist could have been avoided.

Lucy uses soundly based psychological principles in her classes. Key to her approach is desensitization and anxiety reduction. She shows parents and children a way to familiarize themselves with new food in a playful, nonjudgmental and supportive manner. She gives parents, who are often at the end of their tether with their child's eating, a method for moving on. Some children I saw that day were reluctant even to go to the table to look at the fruit and vegetables, while others were ready to put small amounts of purée or raw vegetables to their lips; all stages were accommodated in the class. Above all, everyone was having fun!

In this book, Lucy has put together ideas from her fun-filled classes for all of us to use in a simple, straightforward way. Have lots of fun and enjoy this book. Go on, have a go – it really does work!

Dr Catherine Dendy
Clinical psychologist (children's feeding specialist)

Introduction

Often you start making choices about your baby's food before they are even born: breast or bottle, feeding on demand or by the clock, when to start weaning and so on. Then when your baby arrives, it soon becomes apparent that they have neither read the baby manuals nor are very happy about the choices that you've made. And so begins the long and winding road to adulthood. Out of all the decisions that you make about your child's future, there can be few that are so emotive, pressurized and influenced – from within and outside the family – than that of how to nourish your child.

The transition from baby to toddler feeding is an amazing journey of discovery, but can also be completely baffling at times. The confusion comes when you have spent months feeding your child every puréed vegetable under the sun to suddenly find they throw all of mother nature's finger foods on the floor. If you are already going through this process or perhaps just approaching weaning, have a troublesome two year old or a plain fussy six year old, this book offers encouragement and advice to guide you through these tricky stages. The range of activities is suitable for children from as young as twelve months up to the age of seven and is designed to build up your confidence to deal with eating struggles, the bane of most parents' lives.

Many of you will already have several children's cookbooks on the shelf and spend time preparing healthy and nutritious things for your child to eat. However, attempting to get the food into their mouths is another challenge altogether and sadly the cookbooks don't tell you how to do this. This is where Mange Tout can help.

I launched Mange Tout in 2005 and there are now classes in Clapham, Wandsworth, Battersea and Oxford, with more planned throughout the

UK, Australia and New Zealand. Each class starts with circle time, where we enjoy games, stories and songs about fruit, vegetables and healthy eating. After that, table time is where we get down to exploring different fruit and vegetables in their raw, cooked and puréed form. Children are never asked specifically to eat, try or taste the food. Instead they learn about colour, shape, texture and smell, get to practise using cutlery, manipulate the food with their hands and perhaps kiss, lick or make teeth marks in the produce. At the end of each session, the children are rewarded with a Pod sticker and get to select a treat from Pod's Magic Box (filled with a variety of dried fruit). There is no pressure to taste the food, but it happens at some point for each child. Every Mange Tout child has ended the course happily eating fruit and vegetables.

I think the key to doing Mange Tout at home is preparing your child for what's to come on their plate. You can do this by letting them help get the meal ready, singing about the ingredients or telling a story about the food they are going to eat. Most importantly, you must let your child become familiar with the food you are going to introduce to them. Handing most children a large head of broccoli is likely to grip them with a fear of having to eat it. However, asking a child if they can feel or smell the broccoli and perhaps chop or break it up for you is a good introduction to this vegetable.

'That's all very well,' I hear you exclaim, 'but how am I supposed to deal with a newborn screaming for a feed, a toddler pulling at my jeans demanding to paint and a five year old practising their recorder while whizzing up a nutritious culinary delight that they will all sit down quietly to eat?'

I certainly appreciate how hectic parenting can be and how stressful it is when children won't eat healthily. That is why I set up the Mange Tout classes. The beauty of Mange Tout is that you can incorporate it into your normal routines, by selecting a few of the activities to occupy your child while you get on with something in the kitchen. Your child will feel that they are contributing to the meal and, at the same time, perhaps they will kiss the cauliflower they are breaking up, peel the orange they have been rolling or, even better, munch on the raw peas they are shelling.

Mange Tout is all about finding fun in food. It offers parents and children the chance to get used to fruit and vegetables away from the stress and frustration of mealtimes. To help children feel confident about new foods, we must start to involve them in the whole process, from planning, selecting and shopping for meals to preparing and eating foods together.

This book will offer simple, easy-to-follow ideas to help you introduce fruit and vegetables to your child as well as providing effective techniques for stress-free, successful mealtimes. These methods are similar to those used in our classes and will help you to recreate the Mange Tout experience in your home.

Mother of Samuel, 3 years old

My son has been involved with Mange Tout from the very beginning. He has gone from not even wanting to touch the fruit and veg to now sometimes trying them in class. But it is at home where there have been the biggest changes: he is now eating, or willing to try, most fruit or veg. I have used the praise (a lot of high-fives), fun and excitement from the classes and we have lots of fun using fruit and veg at home. He now has a real interest in them and loves shopping for, peeling, cutting and opening them all up to see what's inside. I have been amazed in the short period of time how much of a difference it has made. Thanks Lucy and Pod.

Come and Meet Pod

Introduce your child to this book by reading and exploring the following section together. It is important your child befriends Pod because he will play a crucial role in getting your child involved with the songs, games and food activities.

Hello! My name is Pod and this is what I look like. I am a green vegetable called a mange tout (perhaps you have heard me being called a sugar pea or a snow pea?) The French words "mange tout" mean "eat all" and I'm called that because you eat all of me – the peas inside and the green pod that holds them. I love eating fruit and vegetables because they make me feel good and help me to grow strong and stay healthy.

   If you like, I can be your special magical friend and help you to learn all about fruit and vegetables. Whenever you see me in this book, it means that it is time for you to get involved and have some fun. Make sure you invite your friends and family to join in too. The following story tells you about one of my many adventures. Perhaps someone could read it to you? I want to tell you a story about how I helped this little girl called Sasha who kept catching colds.

Sasha and Pod

Sasha lived with her family in a very big house that they shared with lots of other families. Sasha had lots of friends to play with and there was always someone to talk to. But every now and again, Sasha would get a tickly nose, start to cough and sneeze and sometimes her throat would get sore too. Whenever that happened, her mum would say that she had to play by herself until she was better. "I want to go and play with my friends," Sasha would grumble, but her mum always said, "It's just not fair to give all your germs to everyone else." Sasha quite often had to spend a few miserable days by herself until the cold got better. She would sit by the window and watch everyone else playing in the garden and feel very lonely and unhappy.

One afternoon, Sasha sneezed and sneezed and felt so fed up that she started to cry. I heard her crying and popped into her room to see if I could help. "What's the matter, Sasha?" I asked. "How do you know my name?" she asked in surprise. "Well it helps that I am magic," I said, "but I know lots of other things too. How can I help you?"

I talked to Sasha for a while. I asked her what fruit and vegetables she liked and explained how they would help her body to fight the germs that gave her a cold. "Oranges are one of the best for fighting cold germs," I said, "because they have lots of vitamin C." "I don't like oranges," said Sasha, "the skin is very tough and bitter and it hurts my fingers to get the peel off." "Well, what you need are some sunshine smiles, I'll see what I can do!" I winked at Sasha and disappeared.

Sasha looked all around the room, but she couldn't find me anywhere. She was just about to call for her mum when the door opened and in she came. She was carrying a plate with something on it, but Sasha couldn't see what.

"I met my friend Jane, who is a nurse, when I was out shopping. I told her all about you and how many colds you keep getting and she said that you needed some of these. Now I know you don't like oranges, but she said that if you suck them like this you can drink all the juice to help your sore throat, and munch all the fruit to give you the vitamin C to fight the cold germs."

Sasha's mum put the plate on the table and held out a slice of fruit to her. "It looks like a smile," Sasha said and she sucked on the sweet juicy fruit. "You're right," said her mum and smiled, "a plate of sunshine smiles! I hope you like them." And of course she did!

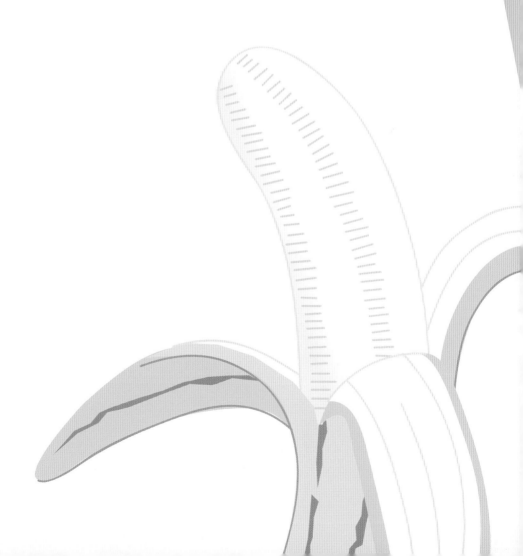

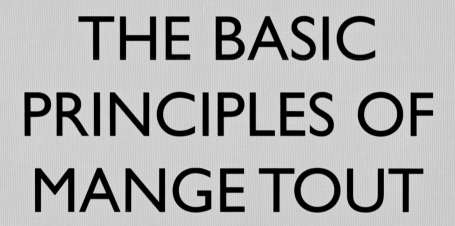

THE BASIC PRINCIPLES OF MANGE TOUT

Mange Tout achieves great success because food is offered to children in a completely different context to that of mealtimes. Children are not presented with food on a plate and expected to start eating. Instead, Mange Tout takes away the fear, stress and pressure sometimes associated with the meal table and eating and instead puts in some excitement, fun and discovery.

Many parents have made great progress with their child's eating by following the Mange Tout guidelines and techniques set out in this book. What's important to remember is that Mange Tout is not just about reading and understanding it for yourself, it is about learning with your child.

By creating opportunities for you and your child to explore and learn together, you will soon begin to discover the fun in food. You may even surprise yourself and encounter something new or perhaps enjoy a food that you thought you'd always hated.

Eight Indispensable Principles

These are the beliefs that are fundamental to Mange Tout. Firstly, we have the golden rule for Mange Tout success:

1) Never ask a child to eat, try or taste anything

At Mange Tout we go back to basics and start at the very beginning, having fun with discovering food and exploring the origin, shape, colour, texture or smell. No one is ever asked to eat, try or taste the produce. Mange Tout offers lots of different ways to experience fruit or vegetables before suggesting that you or your child might kiss, lick or even crunch them.

This may sound strange considering the whole reason that you are embarking on this journey is to get your child eating some good food. However, children do not respond positively to phrases such as:

'Eat it all up.'
'Just try a little bit.'
'But you don't even know what it tastes like.'

Phrases like these can send children into a panic. A child might feel that they can't manage to 'eat it all up' (what if they don't like it?). Sometimes a child may feel threatened by a plea to 'just try a little bit' (why? What is wrong with it? It's obviously horrible if I only have to 'try a little bit'). And sometimes a child is not old enough to understand the complicated logic of reason and consequence when dealing with something new or unusual (I don't need to taste it first because I can see it looks strange/ different). The following story illustrates this point well.

Saskia and the mange tout

One morning at Mange Tout, a new parent came to attend a trial class with her two-year-old daughter. They had arrived early, just as I was pouring the mange tout into a large bowl ready for circle time. The mother's face grimaced as she spotted the green pods and exclaimed, 'You'll never get her to eat those. She hates peas and anything green.'

I greeted Saskia with a huge smile and invited her to come and help me with the green pods. We sat on a brightly coloured mat in the centre of the room and began feeling and counting the pods, trying to guess what was inside them. 'I expect there is treasure inside,' I whispered to Saskia. 'Shall we have a look and see?'

Having built up excitement and anticipation about the treasure to be found inside, we hurriedly popped the pods open and Saskia squealed with delight at the tiny green balls. I counted the peas and instantly popped one into my mouth, exclaimed how sweet and crunchy it was, and then continued with what I was doing. Almost immediately, out the corner of my eye, I caught Saskia popping the peas into her mouth faster than you can say 'Mange Tout!' Saskia was so caught up in the drama and excitement that she'd obviously thrown all green pea grudges out the window and was eagerly popping the next pod whilst mum stood flabbergasted in the corner of the room. Later that morning, Saskia also put the entire pod into her mouth and began crunching it following the 'Mange Tout Are Good for Me' song.

What is important to remember here is that Saskia was not asked to eat the peas; however she responded positively in relaxed company and without any pressure or anxiety to inhibit her.

Children don't respond in the same way to any given situation, therefore it may take you a little more time using the introduction techniques to build your child's confidence. Give positive praise for every small step taken.

2) There is no such thing as a perfect parent

Mange Tout is not about being a perfect parent or leading the perfect healthy lifestyle. There is no such thing! Mange Tout is about taking the pressure off mealtimes.

A guilty conscience is one of the main problems when it comes to making decisions about what to feed your child and when. If your child doesn't want to eat what has been prepared as the family meal, then do not force them. However, avoid giving your child something else as this will only serve to reinforce the idea that every time they refuse a meal, they will be given something different. Your child will not starve after missing a single meal. If you stick to your decision, they will be ready to come and eat at the next mealtime. Offer fruit or vegetable sticks and plain crackers or rice cakes with water in the time between. The following story illustrates the above point well.

My older brother and a plate of shepherd's pie

My parents consciously planned that their children should develop good eating habits by example and from simple rules. They would always offer a well-balanced meal, but try not to make a fuss if it was refused. We all knew that if we ate our main course, then we would be offered a choice of fruit afterwards (grapes and cherries were our favourites). On one particular day, my brother was so anxious to return to his game in the garden that he declined the shepherd's pie and beans – usually a favourite. My parents explained carefully the consequences of his choice, but he declared himself to be 'not hungry' and returned to the garden. After finishing their own meal, they thought it was a shame to waste the small extra portion, so they polished it off between them (this was in those premicrowave days!) About ten minutes later, my brother returned having decided that he had changed his mind and would like to eat his lunch. I don't know who was more shocked, my brother or my parents. My brother spent the afternoon munching on frozen peas (which he grew to love as his favourite snack) and raw carrot sticks, but the lesson was well learnt and he never repeated the experiment.

BUT MY CHILD DOESN'T EAT ANYTHING

If you do find yourself worrying about your child's food intake and want to monitor what they are eating, try not to place too much emphasis on the consumption of one meal. Focus on the day or, even better, the week as a whole, and you may surprise yourself with how much they have actually consumed.

In the first two years, your child's growth and weight increase is astonishing and their appetite may seem insatiable. However at around two years old, a child's appetite will trail off, but this is no cause for concern. On average, most children between the ages of two and five only eat one and a half to two meals per day, so do not expect them to miraculously polish off three full meals every single day.

If we think about it logically, once a child has learnt to walk they soon begin to lose their baby fat and their body becomes leaner due to activity. Their newly proportioned body requires less energy/fuel to function and therefore needs to eat only half the calories per pound of body weight that they did when they were a baby and growing at such a fast rate.

3) Try to recognize your hang-ups about food

In order for Mange Tout and Pod to have a positive influence on you and your family, it is imperative that you take a close look at the food choices you make for yourself every day. Does your diet honestly reflect the range of foods you'd like your child to be sampling? Perhaps there are particular foods that you exclude from your child's diet because they're something you don't enjoy? Children are great imitators regardless of their age and look to parents as role models. Just by setting a good example you are making an excellent start.

OFFER LOTS OF VARIETY

From the onset of weaning, a baby's taste buds are developing and constantly changing at an alarming rate. Between the ages of seven and twelve months, there is a small window of opportunity – usually around sixteen weeks – where it is important you introduce as many different flavours as possible to your child whilst their taste buds are at their most receptive. Keeping your baby's food bland for too long will result in shocked reactions to something as simple as a fresh, tangy tomato sauce. Try adding herbs and spices to food from nine months and remember that many fruit and vegetables make fantastic finger food or great teething tools. Do not worry if your child makes a mess – let them have fun exploring the texture. It's more than likely that they will end up with their hands in their mouth at some point and tasting will take place unconsciously.

Initially young children need to be offered the whole range of flavours and textures available in order to develop their own preferences. Some new tastes will need to be offered as many as ten to fifteen times in order to be accepted and if you stop offering after one refusal, you may be putting a limit on the kinds of foods they will eat as they grow older. Do not give up after the fourth or fifth attempt because acceptance could be just around the corner.

4) Prepare your child for what's to come on their plate

'I'm not going to tell you what it is, but I want you to eat it.'

As a parent, you probably take time to flick through various recipe books and select something interesting and nutritious that might entice your child's appetite. You may also spend a great deal of time preparing the meal to get it right, putting particular effort into the final presentation. At this crucial moment, your hungry child may well respond to the offered food with great suspicion, perhaps refusing to touch or eat any of it. This situation is often the cue for a tired, frustrated parent to feel annoyed, exasperated and disappointed, while the child continues to feel confused, upset or even scared.

SO WHAT CAN I DO NEXT TIME?

It might be helpful to put yourself in your child's place. Imagine that you have been invited to dinner at a friend's house and you are served an unidentifiable and unfamiliar dish. How do you feel? How do you respond? You cannot risk offending your host, so you probably say something like, 'This looks interesting – I don't think I've ever had it. Did you make it yourself?' Most young children have yet to develop such sophisticated language skills to help them cope in a stressful situation. However, regardless of their development, children still experience the same feelings of fear and perhaps loathing in response to something new or strange.

Now, try to picture yourself at a Bedouin desert feast being offered an array of delicacies that could be anything from sheep's eyes to monkeys' brains. Would you be worried and scared and lose your appetite? I'm sure I would. So why do we hold such high expectations of children when it comes to trying and experimenting with new foods?

Cast your mind back to early memories of certain foods you disliked, but were perhaps forced to eat. How many negative associations do you still have with particular foods that prevent you from serving them or perhaps trying them again all these years later? If the foods in question had been presented in a less intimidating way or with more consideration for your preferences, perhaps they would never have caused an issue. A friend of my mum's was forced to eat lumpy mashed potato at school, something she regards as a dreadful experience. To this day, she will not eat potato, regardless of how it's presented to her on the plate. Similarly, I have an enormous aversion to beetroot, which stems from having been sick on a day when I ate it. I was six years old and the sickness was down to a bug circulating in school, however to this day I hold such a dislike for beetroot that I will retch on chewing it. Believe me, I suffered a traumatic time the week we explored beetroot in class.

It is so important that as parents you appreciate how daunting a plateful of vibrant vegetables might be for a toddler. Especially one who doesn't understand, 'But it's good for you!' and who perhaps doesn't recognize it as the orangey-green mush he was gulping down six months ago. Children don't like unexpected surprises on their plates any more than we do. The key to encouraging them to love fruit and vegetables is to make sure they are prepared and ready for something new.

5) Involve your child in the whole process

To help children feel confident about new foods, we must involve them in the whole process from planning, selecting and shopping for meals to preparing and eating or sampling foods together. On a rainy day, pop to the supermarket and spend time in the fruit and vegetables section, allowing your child to touch the produce, ask questions and point out their favourite colour. Choose something new to take home to explore and experiment with. Imagine the pleasure of buying a new and unusual fruit or vegetable and deciding with your child:

> *What it might smell like.*
> *How you're going to prepare it.*
> *Whether it's sweet or sour.*
> *Who will kiss it first.*

Work with your child when trying new food. Explain that even you find certain foods tricky to taste and enlist their help. Make a chart for both of you (or the whole family) and see who can eat five fruit and vegetables a day and get five stars.

It's important that we teach children how and where food grows, why we need good food inside us and how particular foods can do wonderful things for certain parts of our body. The chapter 'Doing Mange Tout at Home' will help you introduce fruit and vegetables to your child while having fun and learning together.

6) Get a little messy

Some children can be anxious about touching something with their hands, for example a slice of wet, slippery tomato. Or they can worry that something simple like squeezing an orange to make their own juice to drink will leave their hands or clothes all dirty. Our often negative attitudes towards dirt and messy play, along with constant reminders to children to 'stop getting dirty', are affecting the way children learn.

Research suggests that a third of children avoid certain types of messy play because they are worried about dirt. Children learn through play and, best of all, when they are actively involved, which may well cause a bit of a mess. We need to provide the opportunity for children to take part in some messy play, from cooking and art and craft to gardening, and let them know that it is OK to get dirty.

7) Even the longest journey has to begin with a few small steps

Always allow your child to watch you carry out an activity and do not force them to join in. Watching you will allow their confidence to build and once they realize you are having fun, they will want to join in too. Remember that Mange Tout is always a work in progress and that getting your child to read, play, talk about and explore fruit and vegetables inside a daily routine will increase their knowledge and confidence along with gradually building acceptance. Try incorporating the following suggestions into your everyday activities and you'll soon see changes in their attitude to food:

- Talk to your child openly about food. Ask them their opinion when selecting fruit and vegetables. Explain how it helps our bodies to work and stay healthy.
- Involve your child in the whole process, from choosing and buying to preparing and eating (remember that many vegetables are better for us if eaten raw or only slightly cooked).
- Take time to look at fruit and vegetables outside of normal mealtimes. This book offers guidance and suggestions for suitable activities.

Above all, do not worry. Small changes will make a noticeable difference very quickly and even the longest journey has to begin with a few small steps.

8) Offer positive experiences

It's important for parents to encourage their child to be confident and adventurous with food. Children need the opportunity to get their hands on food in a relaxed and fun setting and, above all, they need to be praised for their involvement, even if only a small step is taken.

LITTLE THINGS MEAN A LOT

Children show an amazingly speedy reaction to positive instruction and it is the most successful way of achieving your goals with your child. Walk into a room of chatting children and ask them to be quiet and you will probably be very disappointed with the response. Walk into the same room and quietly single out one child for praise because they are sitting quietly and within seconds the whole room will follow suit.

We are aware that children need direction and guidance, but all too often we find ourselves being negative, even though we know that words of criticism can hurt and stay with us for a long time. Remaining positive can be difficult, especially during times of stress, but the results achieved are so spectacular that you will want to make it a permanent part of your life with your child. The first step is to train yourself to look for something to praise. To begin with this may not be easy, but it soon becomes second nature. Part of the trick is not to make a song and dance about it, but quietly to say something like, 'It was really good of you to get dressed without any help today,' or 'It's lovely when you enjoy the dinner that I've made for you.'

YOU CAN MAKE A DOG WAG ITS TAIL WITH JUST THE TONE OF YOUR VOICE

If you imagine your child's feelings are his tail, would that tail wag very often when you speak?

If you are brave, you could try something that has been undertaken in a few schools by very courageous teachers. Simply leave a tape recorder in a room to record your interactions with your family. Let the tape run to the end so that you forget that it is there. Later listen to it on your own to discover how many times you say 'No!' or make a negative comment. You might be very surprised to hear that you are not as positive as you thought.

BEING POSITIVE DOESN'T MEAN YOU CAN NEVER SAY NO

It doesn't mean that you have to smile all the time either, but it does involve a change of attitude and allowing children to make some choices for themselves. For example, your child may ask, 'Please can we have chips with tea tonight mum?' Your reply could be something like, 'Well, I hadn't planned to have them until the weekend. I thought we might have pasta tonight, but we could swap over. What do you think? We could wash and prepare the potatoes together.' In this way you only have to make a slight adjustment to the menu and your child gets to feel valued and can make a decision about when to have the chips.

Mother of Christian, 2½ years old

Christian had been on chemotherapy for ten months so I was desperate for him to get some "good stuff" inside him. This was proving to be a bit of a battle until he started Mange Tout. Whilst he may not be a spinach-eating toddler, food no longer scares him and mealtimes are positively joyful!

DOING
MANGE TOUT
AT HOME

Mange Tout is all about providing positive experiences. The ideas in this chapter will help to support you by offering activities, games and learning experiences that are flexible, that you can incorporate into your daily routine and that will, in turn, complement your parenting.

In the Mange Tout classes, we look at two vegetables and one fruit in the space of forty-five minutes. We begin with circle time, introducing the produce with songs, then move onto table time and hands-on exploring of the food. We end up with more circle time, playing games related to the produce. Parents can then introduce the same three items throughout the week at home using the ideas from the class.

Depending on the age and ability of your child, you may decide to select just one item and spend only five minutes exploring it with them. You could then try something different with it over the course of a few days. Or perhaps you might like to organize a group of friends and decide on one item, for example tomatoes, and spend twenty minutes following the tomato section from start to finish.

Remember that repetition is one of the keys to success and this may mean offering food ten to fifteen times. Use the games and songs to your advantage to make the repeated fruit and vegetable even more exciting the next time round.

Before you start

The following section describes briefly the three stages of Mange Tout. It will also help introduce you to the methods we will be using throughout this book and is a good place to start before going on to the individual fruit and vegetables. Depending on how averse your child is when it comes to fresh produce, a gradual introduction using pictures may work well. Perhaps your child responds positively in different environments? In that case, the supermarket fruit aisle would be a great place to start. Alternatively, a simple song with actions could be the perfect place to begin your fruit and vegetables fun.

INTRODUCTION

- Use simple picture books or wall charts to look at and talk about fruit and vegetables with your child. Ask how many they know or recognize and where they grow. Discuss your favourite shapes and colours.
- Visit a fruit or vegetable farm, a farmer's market, greengrocers or supermarket vegetable aisle. Point out all the different varieties of a particular produce. See if there are any you don't recognize. Choose something new to take home and explore (do not talk about eating it).

INTERACTION

The following activities must be done away from the table that is used for mealtimes.

- Set up a tablecloth on a floor, in the garden or use a different room and table.
- Handle the produce in its raw and, if possible, field-fresh state. This can involve putting it away into the fridge or fruit bowl; letting your child wash the produce in a bowl of water; hiding and then hunting for the fruit or vegetable; singing and dancing with the produce; holding, stroking, rolling, catching, passing, peeling, breaking/snapping, tearing and squeezing.
- Discuss the different senses we use to explore. What does it feel like? Is it bumpy or smooth? Is it cold on your face? Is it hard or soft? Can you smell it? Can you kiss it? Can you lick it? Can you brush your teeth with it? Does it crunch? Can you see your teeth marks in it?

BUILD ON YOUR PROGRESS

- Involve your child in cooking or simple gardening. They may then want to eat the peas that they have grown, shelled and prepared.
- Offer choice. Your child may not have liked cooked carrots, but she might eat them raw and grated.
- Repetition is the key to continual success. Keep up the action songs and funny games even following your successes.
- Create star or sticker charts for the whole family to track progress.
- Use positive praise throughout every experience.

Mother of Max, 3 years old

When we started at Mange Tout, Max was very good at eating a variety of fruits, but refused all vegetables. At best he would ignore them and at worst he would throw them off his plate. The only way I could get him to eat vegetables was by grating and cooking blandly coloured vegetables and adding them to a cheese sauce or, if I was lucky, sprinkling some brighter colours on to a home-made pizza.

Now, Max is happy for vegetables to be on his plate. He will handle the vegetables, putting them on his fork and in his mouth and occasionally eating a bit. In fact, every time he sees a cauliflower, he wants to kiss it! He watches the family intently while we eat our meals and even I am eating carrots to demonstrate that they are fine (carrots are a food I have never liked, so perhaps Mange Tout is helping me as well).

I feel that we are taking a positive step forward after a time when we seemed to be regressing, with something else refused each week. The problem was wider than just vegetables, but now Max is more open to all foods.

Mange Tout is effective and because it involves the children, alongside their parents, they are more interested. It is simple, unpressured and the tips and ideas are great. Max loves raisins, but I had not thought about other dried fruits like cranberries (or 'red raisins' as I call them) – he loves them now. I would recommend Mange Tout to anyone. It is a fun-tastic activity for any child.

The importance of raw vegetables

Something that surprises the parents who attend Mange Tout classes, and indeed perhaps yourself on reading this book, is the range of vegetables that we can enjoy raw. You may even find yourself amazed at how much tastier vegetables are when raw, particularly the ones you never imagined could pass your lips without being cooked.

My earliest fond memories of raw vegetables came about from spending Sunday afternoons with my brothers and my dad on his allotment. Aside from regularly feeding our dad's prized produce to some rabbits in a hutch three plots down from us, our other allotment escapades included outdoor cooking. Supervised by dad and my older brother, we were allowed to chop up and cook our own choice of veggies in an old tin can on top of a smoky peat and compost fire. Into the can would go the water along with carrots, peas, cabbage and even the odd Brussels sprout. However, the water would never reach boiling point and restless and impatient, we would all huddle around the can fishing out pieces of warm, crunchy, almost raw vegetables.

Mange Tout seeks challenges in that we don't always choose familiar fruit and vegetables as our weekly produce in class. We introduce children to a wide and varied selection, some of which they may not have seen before, let alone actually tasted. The fact that our methods work well with challenging fruit and vegetables shows how powerful these techniques are.

If your child has only ever sampled unripe melon or overcooked broccoli, their dislike for the food may be due to a poor seasonal selection or overcooking. I once heard a mother say that the difference between a good peach and a bad one was like comparing heaven and hell!

Brussels sprouts

First up, we won't go for the easy option of mashed or sweet potato, but the most hated vegetable in Britain. Humble, much despised, often seen as a 'thank God it only appears once a year' food, yes it's the 'Christmas vegetable', the Brussels sprout.

As mentioned earlier in the book, your own thoughts and feelings about certain foods and why you don't like them can, without you realizing it, negatively influence your child. A good example of this was at a class held in the final week of term before Christmas. The produce I had selected were cranberries, mange tout and Brussels sprouts. I placed a notice on the door to the classroom asking parents not to make any negative sounds or gestures regarding the food so that they didn't influence the children in any way. The parents entered the room with raised eyebrows wondering what on earth was going on. However their faces soon fell as they spied the sprouts, their expressions changing from that of curiosity to dread and disgust. I could almost hear them silently pleading with me to excuse them from the tasting session, 'Please Lucy, don't make us eat Brussels sprouts, we'll do anything!'

Nevertheless, the children got stuck straight in to handling the sprouts and were interested to hear what this strange miniature cabbage was called. Speaking with the parents later, it was quite clear that most of them had not even thought of offering Brussels sprouts to their child because they were regarded as pungent, ghastly things to be avoided at all costs. This was a clear example of how perceptions about something you dislike immediately impose restrictions on what your child will taste.

Within five minutes of the class starting, children were peeling and devouring the fresh crisp leaves of raw Brussels sprouts, much to the amazement of the parents. To put the icing on the cake – or perhaps the

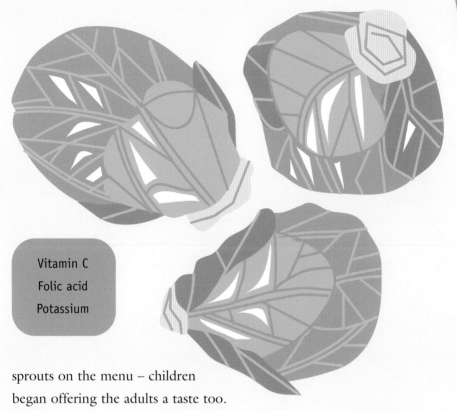

Vitamin C
Folic acid
Potassium

sprouts on the menu – children began offering the adults a taste too.

Keeping to the Mange Tout philosophy that anyone who accompanies a child must also participate in tasting, it left parents little room to manoeuvre as tiny hands thrust sprout leaves into unsuspecting mouths, 'Here you try some mummy.' As a result, the majority of parents were surprised at how good they tasted. The remainder of the group agreed that they weren't actually as bad as they thought they'd be.

For healthy skin and a good immune function, think about including Brussels sprouts in your diet.

Fibre-rich Brussels sprouts also help establish a healthier colon and provide protection against rheumatoid arthritis.

How to introduce your child to sprouts

Include your child in the preparation of Brussels sprouts. It's wonderful if you can buy them on the stalk as children have great fun pulling them off and enjoy the challenge of counting how many they have in their bowl.

- Sit down with your child and show how you can peel off the outer layers of leaves. See how far you can peel them and count the layers or leaves.
- Encourage your child to talk about what you are doing together by asking them about the colour, shape, smell and texture of the sprouts.
- See if you can help them to make flower shapes using the leaves as petals.
- Make the activity funny by seeing if a leaf will stick on your nose, cheek or chin. Stick a tiny piece on your tongue and see if they would like to copy (give them a hand mirror so they can see it on their tongue).
- Ask if they would rather hold it between their lips or make some teeth marks in it and see who has the scariest bite.

Don't worry if your child finds some of these tasks too demanding or if they simply refuse. Continue with the exercise and let them watch. Watching you will help reassure them and perhaps they will get involved next time.

Sing a song (to the tune of 'Old MacDonald Had a Farm')

March around clapping hands and rubbing your tummy on the word 'yum' or simply sing whilst peeling the sprouts.

Brussels sprouts are good for me E-I-E-I-O
And so I eat them happily E-I-E-I-O
With a yum yum here and a yum yum there
Here a yum, there a yum, everywhere a yum yum
Brussels sprouts are good for me
And so I eat them happily.

WHAT TO DO WITH BRUSSELS SPROUTS

- Do not overcook them. Overcooking drains the nutrients from the sprouts leaving them with a rather pungent smell that is not particularly enticing. Score a small but deep cross on the base of the sprouts to help them to cook evenly.
- Alternatively, shred the sprouts and sauté with olive oil and garlic.
- Try slicing sprouts raw and add to a home-made coleslaw with grated carrot, onion and white cabbage.
- Combine quartered and cooked Brussels sprouts with sliced red onion and a mild-tasting cheese such as a goat's cheese or feta.
- Toss cooked sprouts with olive oil and balsamic vinegar for an exceptionally healthy, delicious side dish.

Grapefruit

> High vitamin C content
>
> A rich source of bioflavonoids

Go shopping and get your child to look out for all the different sorts of yellow fruit and vegetables. Talk about their size and shapes. Discuss how such fruit and vegetables might be eaten and at what meal. Look to see if there is more than one kind of grapefruit (pink ones are often sweeter).

> Vitamin C in grapefruit is enormously important in combating infection and plays a major part in helping the body absorb iron from other foods.
>
> The bioflavonoids contained in the pith and segment walls strengthen the walls of our tiny blood capillaries.

Sing a song (to the tune of 'Here We Go Round the Mulberry Bush')

Here we go round the grapefruit tree
The grapefruit tree, the grapefruit tree
Here we go round the grapefruit tree on a bright and sunny morning.

This is the way we climb the tree …
This is the way we pick the fruit …
This is the way we squeeze the fruit …
This is the way we drink the juice …

Games to play with family and/or friends

- Roll the grapefruit down a slope.
- Balance one on your foot.
- Play hide-and-seek with it.
- Roll it to one another.

A song with actions (sing to the tune of 'I Had a Little Nut Tree')

Number rhymes and songs like this build numerical skills as well as introducing rhyme and rhythm to your child.

> *Five juicy grapefruit*
> *Growing on a tree*
> *Someone came and picked one*
> *And ate it for their tea.*

Hold up five fingers and pretend to pick a fruit from each one in turn, putting the finger down when 'picked'.

Substitute the name of your child, mummy, daddy or a friend for 'someone'.

An activity to do together

- Draw round your own or your child's hand and wrist.
- Colour it in to look like a tree. Put small pieces of grapefruit on the end of each finger.
- Sing the song above together and eat a piece of fruit each time. Start with the adult taking the first piece. Don't worry if your child doesn't want to eat it. Keep the game relaxed and fun.

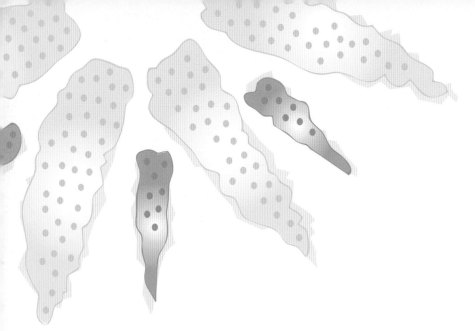

Having fun tasting grapefruit

- Cut the fruit in half and look at the pattern inside.
- Let your child squeeze the juice into a cup. Add fizzy or still mineral water to make a drink. Offer a straw for tasting (perhaps one each as you will want to taste too).
- Stick your fingers in the squashed fruit and lick them.
- Cut the fruit into small slices for sucking.

WHAT ELSE CAN YOU DO WITH A GRAPEFRUIT?

- Make it fun for children to eat by slicing a grapefruit in half around its middle. Use a sharp knife to separate the fruit from the segment walls so that children can scoop the flesh out easily with a spoon.
- Score the skin and let your child peel the fruit.
- Use the peel to make faces or use modelling matchsticks to make animals or monsters from the peel.
- Rub or grate the skin to smell the oil.
- When making muffins, add the juice and grate the zest of a grapefruit into the mixture for an alternative breakfast snack.

Beans and pulses

Beans are an amazing superfood. We can relax knowing that a serving of low-salt and low-sugar baked beans on wholemeal toast is an excellent, nourishing meal. However, in order to avoid just offering this, it is essential to experiment with some alternatives. Whether or not your child is partial to baked beans, it's important to make the transition to another variety as interesting and engaging as possible.

Protein (beans contain almost as much protein, weight for weight, as a good piece of steak)

Magnesium

Fibre

The fibre in beans and pulses helps take care of your heart and circulation and reduces the risk of constipation.

The high level of complex carbohydrates in beans is the best form of energy for active kids.

Activities using dried beans

If your child is old enough to understand the concept of playing with and not eating dried beans, try introducing one of the following activities. Smaller children can also get involved under strict adult supervision, but you should never leave a child unattended when dried beans are within reach. Beans and pulses are unbelievably cheap and most local supermarkets stock a good range. Look out for pinto, flageolet, black-eyed, mung and adzuki beans and various lentils.

- Empty two packets of lentils onto a large, clean tea tray. Ask your child to choose their favourite cars and tipper trucks and let them experiment with the weight and texture of the lentils while using their toys to drive paths through the pulses.
- Empty a selection of beans into a large bowl and, using paper and PVA craft glue, have a go at making collages and pictures.
- Play a game of sorting. Provide your child with four small bowls and help them to sort the beans into various colours. Then put them in order of size and, finally, decide which are your favourite beans.
- Help your child to fill a dry, empty plastic bottle with a choice of beans (do this over a large bowl to collect any that do not make it into the bottle). Tape the lid on firmly and use as a shaker to accompany the bean songs.

A rhyme to speak

Mummy is a jumping bean
The biggest bean you've ever seen
She can jump so high
She can nearly reach the sky!

Repeat the rhyme changing 'mummy' to the name of your child.
Use your bean shakers.

A chant with actions

Jumping beans, jumping beans *(jump up and down)*
Jumping in my pot *(make a circle with arms)*
Lots of different colours *(twirl around)*
I could eat the lot. *(mime eating)*

Exploring broad beans

- When available, buy broad beans in their pods and get your child to help you shell them into a bowl. Talk about how they feel and see who can find the biggest bean. How many beans are there in one pod? What does the inside of the pod feel like?

- During this activity, your child may be curious enough to pop one in their mouth or you could initiate this first. They are delicious raw.

- Try blanching broad beans and letting them cool so that their skins become loose and baggy. Demonstrate to your child how to take their coats/jackets off by pinching a tear in the skin with your nail and squeezing the tender green bean out. Feel how slippery it is on your lips and tongue. Is it soft or crunchy? Does sucking make it any smaller?

Kiwi fruit

Kiwi is the best source of vitamin C available in fruit, containing twice as much as the same amount of oranges. And kiwis also contain more fibre than an apple.

An excellent, extremely gentle laxative that is ideal for youngsters who are often constipated.

This is a very strange looking little fruit and not immediately appetizing. Point them out to your child in the shops and discuss whether they are a fruit or vegetable. You could also talk about whether they are sweet or sour, if they need cooking and what colour they are inside.

Vitamin C
Fibre
Beta carotene

If you are able to handle the fruit, squeeze them gently. The softer they are, the riper and sweeter they will be. Get your child to select a few to experiment with at home. Say something like, 'Shall we take a few home to find out what they are like?' In this way, your child will know that they are just for discovering and experimenting with rather than worrying if they will have to eat them. Explain that kiwi fruit have lots of vitamin C, which helps fight cold germs, so they are very good for our bodies.

Fun ways to discover/explore/touch and maybe taste

- The skin of a kiwi is unlike most other fruit and feels almost furry. But surprisingly, it is thin and can be eaten as long as the fruit has been well washed (although it's not to everyone's taste).
- Decide whether to cut the fruit lengthwise or crosswise – each will give a different pattern. At one end of the fruit, beneath the skin, is a sharp little spike. Show this to your child and feel how hard and sharp it is. This should be removed before eating the fruit unsupervised.

- The fruit is translucent and easily chopped by your child with a blunt knife. It feels slippery and fingers will get very sticky, but licking them clean may well be the first tasting experience.
- Use a lemon squeezer to see how much juice comes out of one half.
- Encourage eating by placing the kiwi fruit in an egg cup and slicing the top off to eat it like a boiled egg.

Sing a song (to the tune of 'London's Burning')

Kiwi furry, Kiwi furry
Green and juicy, green and juicy
Sweet, sweet! Sweet, sweet!
Keep me healthy! Keep me healthy!

Games to play with family and/or friends

- Roll the fruit. Do they roll in a straight line? How are they different to a ball?
- Cut the fruit into thick slices and stack them to see who can build the highest tower.
- Carry a fruit in a dessertspoon.
- Hide some around the room and let your child collect them in a bag or basket.
- Make up an eating game. If your child is happy to eat a small piece of fruit, scoop out the flesh from half a kiwi and dice it into small chunks. Keep the shell of the fruit to hide the pieces underneath. Take turns to roll dice to determine how many pieces of fruit you can have to eat.

Some more activities using kiwi fruit

- Use an old toothbrush to gently scrub the skin clean.
- Slice thinly and lay over the end of a finger to see if you can see through it.
- Draw a circle, let your child colour it green and then dot it with black for the seeds. Look at the fruit to see how the seeds are arranged.
- Use the whole fruit as a body and add limbs made from other fruit or vegetables.
- Use the unpeeled fruit as a head and add seeds or pasta shapes to make features (like Mr Potato Head).
- Kiwi fruit grow on trees trained into a tunnel. You could make a play tunnel with an old blanket or sheet and stick pretend kiwi fruit inside with Sellotape.

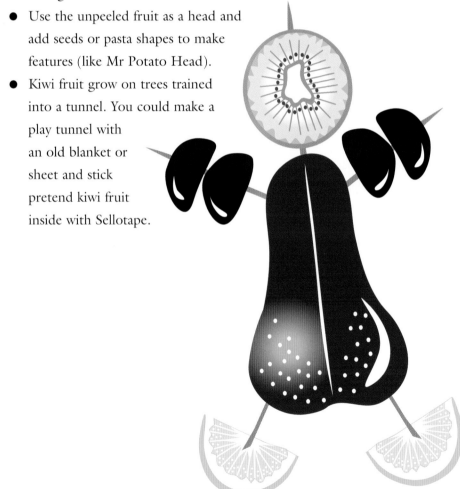

Mushrooms

Mushrooms are something that I didn't grow to love and enjoy until I went to university. All of a sudden, I was throwing them into stir-fries, slicing them into omelettes and making a mean Stroganoff. Cooked mushrooms do have rather a strong flavour, one which many people believe may be too sophisticated for young children. However, once again we proved it not to be the case at Mange Tout as children munched their way through the raw and cooked fungi. Don't let the strong and distinctive flavour of mushrooms deter you from offering them to your child – they may well surprise you.

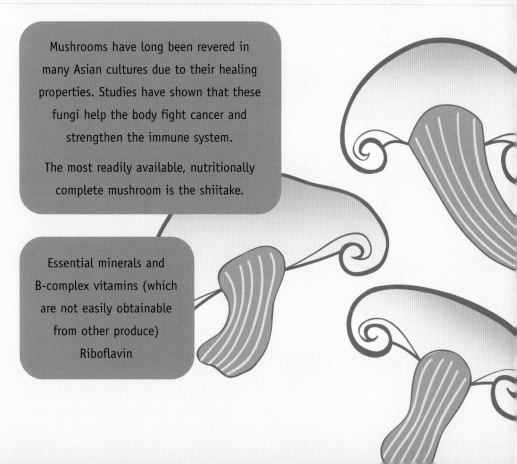

Mushrooms have long been revered in many Asian cultures due to their healing properties. Studies have shown that these fungi help the body fight cancer and strengthen the immune system.

The most readily available, nutritionally complete mushroom is the shiitake.

Essential minerals and B-complex vitamins (which are not easily obtainable from other produce) Riboflavin

Activity time

- At the supermarket, remember to point out the huge variety of mushrooms that are available to buy. Talk about the different names and the shapes or colours of them all. Decide which you would like to take home and play hide-and-seek with.

- At home, hide the mushrooms round the room and get your child to go hunting for them with a small box or basket. How many did you find? Can you line them up in order of size?

- Show your child how you can pull the stalk out of a mushroom to reveal all the brown gills hidden underneath the umbrella-shaped vegetable. Smell the stalk to see if it is different to the rest of the mushroom. Stroke the velvety brown underside very gently with your fingers. Is it rough or smooth and does it feel the same on your tongue – perhaps it tickles when you lick it? Peel the skin off the mushroom to see the colour underneath. Once you have peeled a mushroom, give it a good squeeze. It's amazing how much juice comes out. What colour is the juice?

- Give your child a blunt or plastic knife for slicing the mushroom and remember to praise them for their excellent work.

- Show how you can make teeth marks in the white part of the mushroom.

- Do some mushroom printing with the mushroom stalks, round heads and umbrella-shaped slices using some beetroot or strawberry juice. Even water works well on coloured paper.

Sing a song (to the tune of 'Twinkle, Twinkle Little Star')

Pod says smell it
Pod says kiss it
Pod says lick it
Does it crunch?

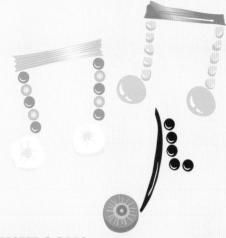

WHAT TO DO WITH MUSHROOMS

- Be aware that mushrooms absorb oil like a sponge so use sparingly if you are frying or add a dash of water or soy sauce to get them cooking.
- Add raw to salads or offer raw with a healthy dip. The flavour of raw mushrooms is a little subtler than cooked and there is a higher nutritional content.
- Great tossed into stews and soups as a fat-free way of adding extra flavour.
- Mushrooms give a good meatiness to vegetarian dishes.

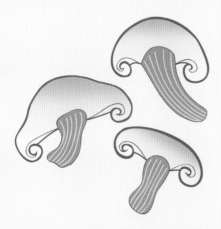

Make sure you talk to your child about the importance of not picking or eating wild mushrooms and fungi that are growing outside.

Celery

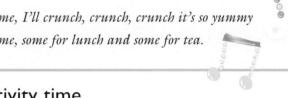

> Potassium
> Vitamin C
> Fibre

This cheap and versatile vegetable used to be among my least favourites. In fact I would go as far as to say that before I started Mange Tout, I would go out of my way to avoid eating it, picking celery out of every salad or stew and wincing at my mother who would happily crunch away on sticks of the stuff.

Teaching Mange Tout has helped me to accept celery, although I have to admit that the early days of celery crunching left a bitter taste in my mouth (one that I had to conceal from the children!). My other confession is that I would still rarely choose to crunch my way through a stack of celery sticks, but I can be tempted with a big pot of hummus. I also have a weakness for a good Waldorf Twist – celery, apple, lime and ginger juiced together and served over ice – delicious!

Sing a song (to the tune of 'Old MacDonald Had a Farm')

Celery is good for me, I'll crunch, crunch, crunch it's so yummy
Celery is good for me, some for lunch and some for tea.

Activity time

- Hold a bunch of celery and get your child to pull a stick off and listen to the sound it makes (a very satisfying snapping sound if you have a fresh celery bunch). Snap the celery in half and see how the stringy bits separate themselves. Peel them off and feel the juice spray on your face.
- Starting at one end of the celery stick, crunch along making a long line of teeth marks. How loud can you make your crunch? How many teeth marks can you make?

A song with actions (sing to the tune of 'The Hokey Cokey')

A really fun way to get kids familiar with celery and something that helped put an amusing slant on an unloved vegetable for me was 'The Hokey Cokey Celery'. Stick with the original actions, only do them holding a stick of celery!

You put the celery in, the celery out
In out, in out, you shake it all about
You wave the celery and you turn around
That's what it's all about
All together now
Whoa-o CELERY
Whoa-o CELERY
Whoa-o CELERY
Knees bend, arms stretched
CELERY!

> Celery is a rich source of folate, which makes it an excellent addition to salads for women planning pregnancy.
>
> The essential oil from the celery seed has a powerful calming effect on the nervous system.

WHAT TO DO WITH CELERY

- Celery is a fantastic way of introducing lots of different flavours to your child. Dip celery sticks into a variety of purées and dips.
- It is also excellent for those troublesome months of teething. Leaving the celery stick long enough to hold on to, peel off the stringy layer using a potato peeler and place in the freezer or fridge. Use as needed (remember never to leave a child unattended with food).
- The great crunchy texture and mild flavour of celery makes it a great snack to nibble on with or without a dip.
- Juice along with apples and carrots for an added healthy zing.
- The nutritional benefits make it a good addition to winter soups.

Tomatoes

Refer to the chapter on growing your own fruit and vegetables to see how easy it is for even the not so green-fingered novice to grow some tomatoes out of a growbag.

Tomatoes are extremely rich in antioxidants such as carotenoids and lycopenes, making them good protectors of the cardiovascular system and effective against some forms of cancer.

Vitamin C
Potassium
A good source of carotenoids

A song with actions (sing to the tune of 'Frère Jacques'/'Are You Sleeping?')

Start the song by getting yourself into a small ball, crouching down low and wrapping your arms around you (children are particularly good at doing this). Pretend you are the tomato.

> *Red tomato, red tomato*
> *On the vine, on the vine* (arms outstretched and jiggle them)
> *Ripe and sweet and juicy, ripe and sweet and juicy* (rub your tummy)
> *Please be mine, please be mine.* (nod head and point to yourself)

Activity time

Buy a range of fresh tomatoes – cherry, plum, pomodorino, yellow and green varieties – do they feel or smell different to each other? Let your child have them to play with in some small bowls or with their tea or cooking play set.

You can sing

Buy some cherry tomatoes on the vine and pick and count them off the vine. At the same time sing (to the tune of 'Ten Green Bottles'):

Ten red tomatoes growing on a vine, Lucy came and picked one, this tomato's mine!

(Alternate between you and your child until they have all been picked – then count how many each person has.)

Fun ways to discover/explore/touch and maybe taste

- Hide the tomatoes around the garden or kitchen and go hunting for them. Have fun washing them in a small bowl of water while talking about the texture of the skin. Give the tomato a kiss to see if it's bumpy or smooth. Can you feel it with your tongue? Try licking the water off the tomato.
- Offer a small plastic serrated knife to slice the tomatoes open with. Show how you can push your finger inside and scoop out the seeds onto a plate or bowl. Show how noisily you can slurp the seeds out by sucking on half of the tomato.
- Paint your lips with the juice. Demonstrate how shiny and beautiful your lips look with tomato juice on them.
- See how many slippery seeds you can count. Can you try and pick one up? Does it stick on your tongue?
- Let your child put some tomatoes in a bowl in the centre of the table for everyone to enjoy before or during a meal. This is particularly exciting if they are home-grown ones.

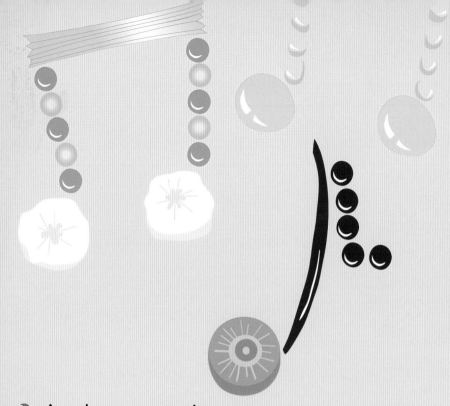

Another song to sing (to the tune of 'Twinkle, Twinkle Little Star')

Try boiling some large tomatoes until the skins split. Once cooled, have a go at peeling the skins off with your child. Once the tomatoes have been peeled it is a great time to use the following song:

Pod says smell it (enthusiastic smelling to encourage your child to copy)
Pod says kiss it (lots of vigorous kissing)
Pod says lick it (lick with approving sounds of how delicious it is)
Does it crunch? (an enormous crunch – if your child is hesitant, then
    show how you can do a baby crunch or even just some teeth marks
    in the tomato)

TOMATO KETCHUP

Tomatoes are quite acidic, but ketchup tastes sweet, which is why so many children will happily pour rivers of tomato sauce onto their plate, yet turn their noses up at a beautiful, fresh, ripe tomato. The fact that children might even be willing to eat ketchup straight from the bottle is because the added salt and sugar make it so tasty.

However, I will go as far as saying that providing you buy a low salt, reduced-sugar variety (I would also choose organic over regular brands), a tablespoon of ketchup three times a week can actually be quite beneficial to your child's diet. Make sure your child understands the restrictions that you place on having tomato sauce with a meal. You can even make a chart with sticky red dots to record the use of it so that your child can understand it in a visual sense. This can be particularly helpful is you are struggling to limit the use of sauce at mealtimes.

WAYS TO INCORPORATE TOMATOES INTO YOUR FAMILY'S DIET

- Use passata (sieved tomatoes, available in all supermarkets) as an alternative to ketchup.
- Try stuffing large beef tomatoes with a favourite rice or pasta dish to make an edible bowl.
- Offer fresh salsa as an alternative to ketchup (see recipe on page 155). Delicious with breadsticks and soft cheese.
- Tomato purée and passata are great on pizza and in pasta sauces. Add sun-blush tomatoes as a great extra topping for a healthy home-made pizza.
- Blunt the ends of some wooden skewers and slide on tomato halves, pieces of cheese and some green grapes.
- Roasting tomatoes reduces the acidity and brings out the intensity of their natural sweetness.
- Try offering sun-dried tomatoes (although they can be a little tough to chew) or much softer sun-blush tomatoes. Even though both varieties tend to be soaked in olive oil, drain them and dab with kitchen roll. Cut into small pieces and serve in a small dish with some raisins.

Carrots

I could not believe my eyes when a parent arrived at Mange Tout and grimaced at the sight of … carrots! I thought it was going to be a straightforward week with what I'd assumed were popular favourites, but to my amazement there were many parents and children who were not in the least bit fond of this vibrant orange vegetable.

At the shops, look out for the different varieties of carrots. Chanteray carrots (they look a bit like small orange spinning tops) and baby carrots are very cute and may be enticing for small children. It is even possible to get purple carrots (they tend to be a rather brownish purple colour, not unlike long, thin beetroots). Many farmers' markets and some supermarkets sell bunched carrots with the fine, feathery stalks still attached.

Betacarotene
Vitamins A and C
Potassium

Vitamin A is important for good eye health and helps the body fight infection, so although carrots don't really help us to see in the dark, they do keep our eyes functioning well.

Activity time

If you are feeling adventurous and don't mind a bit of mess in the garden, fill a pot or bucket with some compost and bury the bunched carrots in the soil so that their green stalks are hanging out at the top. This is a great visual exercise to demonstrate to children how carrots (or other root vegetables) grow. Have a look at the growing section later in the book. Once your carrots are safely in the soil, it is time to pick them.

A song to sing (to the tune of 'Oh My Darling Clementine')

Picked a carrot, picked a carrot (mime pulling a carrot out of the ground)
That was growing in the sun (make a big sunshine by moving your arms in a circular motion)
Then we washed it (action of washing and rubbing hands together)
And we ate it (mime eating)
So we picked another one.

- At the end of each verse, take turns to pull a carrot out of the ground. Have some fun washing the carrots and patting them dry with some kitchen roll or a tea towel.
- Help your child to place the carrots in a line in order of height or how fat or thin they are. Ask them what the carrot smells like. Scratch it with your fingernail and see how much juice sprays out.
- Hide the carrots around the room and go looking for them. Try bouncing and jumping like a rabbit while you hunt for them.

How many different ways can you carry a carrot?

- Hold it under your arm.
- Balance it on your shoulder.
- Tuck it under your chin.
- Carry it between your legs.
- Pretend to be a horse or donkey and carry it in your mouth.

Activities using carrots

- Select some sliced, grated and frappé carrot to make a face with on a plate. (To make carrot frappé, peel a carrot on four sides with a potato peeler until you are left with the frappé and a rectangular carrot.)
- Using raw and cooked carrot batons, get your child to brush their teeth. Demonstrate with some loud and enthusiastic noises – 'arghhh' when brushing back teeth and 'cheese' when brushing the front ones. How vigorously can you brush? Does it turn your teeth orange?
- Make teeth marks and patterns in cooked or raw carrot.
- Make some scary teeth by hanging carrot frappé from your lips. Or perhaps make a long tongue.
- Sprinkle grated carrot onto your plate and listen to the sound it makes. Lick it and see if it sticks on your tongue.
- Ask your child to identify which is raw and which is cooked carrot. Discuss the differences in colour and texture.
- Offer your child a plastic serrated knife and get them to slice cooked baby carrots into circles or slices and the cooked rectangular carrots leftover from making carrot frappe into cubes. Do the cubed carrots have a different flavour to the circle ones?

- You could try having some carrot purée to use as toothpaste to brush your teeth with. Or perhaps spread it between carrot discs so that they stick together to make a sandwich. Crunch your sandwich to make the shape of a smile or half-moon.

WAYS TO ENJOY CARROTS

- Carrot sticks are a colourful addition to any meal. Raw, shredded or sliced carrots can be added to a salad. To make raw carrots easier to chew, briefly steam them.
- Offer raw carrots with a hummus, cheese or avocado dip.
  - Try adding orange juice to the water when boiling carrots to give them a tasty zing.
    - Roast carrots with a little honey on them.
      - The day we were doing carrots at Mange Tout, we were also doing apples. Lots of children enjoyed dipping their carrot batons into apple purée. Carrot and apple juice is absolutely delicious.

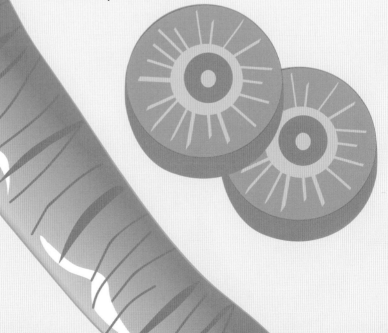

Peas

What is it about peas that make many children find them
so undesirable? Perhaps it's just me, but I find the little
green balls rather cute. They taste so sweet and are immensely satisfying
with mashed potato. However, it seems that quite a few children need a
little more gentle persuasion in the form of fun before peas reach their lips.

First though, let's consider peas from a child's point of view. A pile
of peas rolling around on the plate can seem extremely daunting. When
babies make the transition to solid puréed food, it often gets spat out in

The B vitamins present in peas
are necessary to help convert
food into energy.

They also aid the brain and
nervous system and are great for
healthy skin and cell production.

response to the strange texture, especially when lumpier foods are gradually introduced. Young children find it quite tricky to deal with multiple textures in their mouth. How many times has your child refused yogurt with bits in, bread with seeds or pasta sauce with chunks? Do you ever notice how your child prefers to eat their food separately? Perhaps you've made a forkful of potato for them and buried a pea inside, but they then began to swirl the food around inside their mouth until eventually the pea got pulled out. The same process can occur when a whole spoonful of peas is put into a child's mouth and they feel unable to chew or deal with them all at once.

Many children gag quite easily on cooked peas, however do not be alarmed and moreover do not take this to be a sign of their dislike. A cooked pea separates very easily from its skin once it is in the mouth. The soft, sweet pea is swallowed quite easily, however the slightly tougher skin (especially if overcooked) may not be so palatable and can easily get stuck in the wrong place and be tricky to swallow. A small retching or gagging incident can occur and a child can associate this all too soon with being sick. The natural response is to then dislike peas.

> Remember that Pod is a pea too. Encourage your child by reminding them that Pod would be very proud of them for growing, shelling, smelling, kissing or crunching peas.

A rhyme to teach your child

Use your hand closed in a fist and let your fingers pop out as the peas grow.

Five green peas in a pea pod pressed
One grew, two grew and so did all the rest
They grew and they grew and they just couldn't stop
Until one day the pod went POP!

Activity time

- Buy some peas that are still in their pods and have fun popping them out. If you cannot find podding peas, then mange tout work well too.

- If your child enjoys eating them this way, try lightly steaming mange tout and letting them cool before opening them. Count how many peas are in the pod. Before you pop them open, try and guess how many will be inside.

- With cooked peas, show your child that by gently squeezing a pea, it will pop itself out of its skin. Inside there are two small halves.

- Stick half a pea on your tongue so it looks like you have a green spot on your tongue. Can you make it disappear?

- Remember my story about Saskia and how she got so involved with the activity that she immediately copied me by popping a raw pea into her mouth? You can do the same, but do not ask or force your child to copy you, just exclaim how sweet and crunchy they are and encourage them to choose a baby pea or the biggest one.

- Feel how smooth the pea is by licking it with your tongue. Does it get any smaller if we suck it? Will it disappear?

- Demonstrate the magic of mashed potato. Press a little from your plate on to your fork, add one pea and show how the little pea can stay on the fork even when it is turned upside down.

If your child is happy to explore and not eat, do not worry. It is great that they are getting involved with the activity and not shunning our little green friends. Remember – a chef's love of food comes from hands-on experience, exploring and getting to grips with food itself.

A rhyme with actions

Put some hard frozen peas into a container with a lid on and chant while shaking the container:

Frozen peas, frozen peas
Jumping in my pot
Lots of crunchy green balls
I could eat the lot.

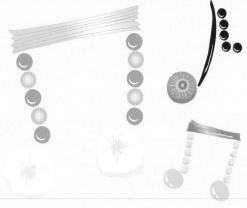

Open up the container and share the crunchy frozen peas. They make a great tasty nutritional snack and because they are frozen, they have a consistent texture and are surprisingly sweet. Many parents were dumbfounded in class to see their children shovelling frozen peas into their mouths faster than you can say Pod. One little boy, who had never eaten a solid vegetable in his life, thought the green balls were brilliant and, much to his mother's amazement, finished off the entire container.

Melanie Campanile, London

I will never forget the day we came home from Mange Tout having experienced raw spinach leaves for the first time. My children dressed up in giraffe and elephant masks and asked me to be their tree. I stood in the kitchen clutching bundles of raw spinach whilst my children played around and munched away at the green leaves. It was hilarious, but highly nutritious!

Oranges

Many children will immediately associate an orange with the idea of a drink, be it orange juice or squash. Sadly however, many of the orange juices and drinks that are readily available in supermarkets and cafés contain added sugar and even nasty chemicals in the form of colorants and artificial flavours. Therefore it is essential that you get your child interested in the real thing.

At the supermarket, have a look at all the different types of orange-coloured fruit and vegetables. Are they all oranges just because they are the colour orange? Point out the different oranges, satsumas, clementines and mandarins, select a few to take home and explore.

Activity time

- Large oranges are great for rolling. Sit opposite your child and roll an orange to and fro.
- Hold an orange underneath your chin and try walking with it.
- Stretch out your arms in front of you with your palms together and face up. Put an orange under your chin, then lift your chin off the orange and let it roll down your arms into your hands. Very tricky!
- Try rolling the orange across the room into a container. Lay an empty box on its side or tape an ice-cream container to the floor for your target.

A game to try

Play a game of orange bowls by placing a large orange in the centre of the room as the jack. Mark a spot to stand behind and see who can gently roll their orange or satsuma so that it lands closest to the jack.

Remember that games give your child a hands-on experience that will build their confidence and acceptance of the fruit. By introducing an element of fun you are creating some positive associations with oranges too. You may even find that your child wants to play a game of bowls or sing a song before they sit down to enjoy an orange. Brilliant – a fantastic portable game and snack.

Vitamins C
and B
Calcium
Potassium

Vitamin C helps the body combat infection and preserves general health.

It also helps the body to absorb iron from other foods.

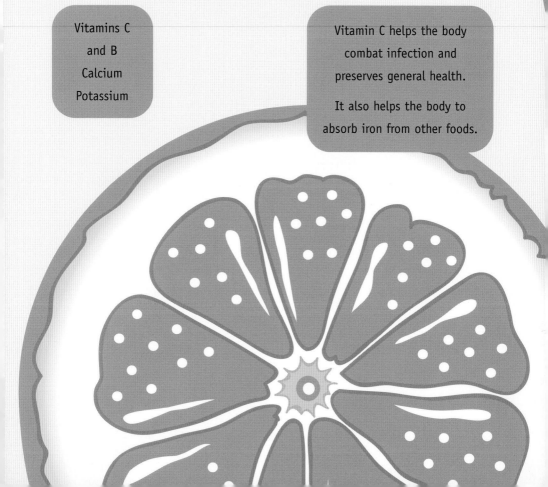

You can sing (to the tune of 'Row, Row, Row the Boat')

Whoever ends up with the orange at the end of the song must be encouraged to feel and smell the fruit before you repeat the song:

Roll, roll, roll the orange
Roll it to and fro
Orange, orange
Orange, orange
Gently as we go.

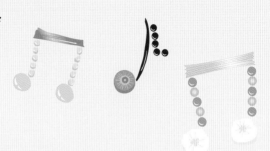

WAYS TO EXPLORE AND ENJOY ORANGES

- To help your child peel a large orange (which can sometimes be quite tough), use a knife to score the orange skin into quarters and slice the top off so they can remove the skin easily.
- Satsumas tend to be the easiest members of the orange family for a child's small hands to peel. Demonstrate how to stick a thumb or finger in the base of the fruit to make a hole to help start the peeling process. Once you have peeled your satsuma, stick it on your thumb like a ring or puppet. Enjoy separating the segments and counting how many there are. Make a pattern with them – a flower, face or even a ladder.

Sing a song (to the tune of 'I Had a Little Nut Tree')

Draw round your hand on a piece of paper and place a segment on the top of each finger. At the end of each verse, the person called can choose a piece of orange. Do not force your child to eat it. There are many activities to follow that may encourage them.

Ten juicy oranges growing on a tree
(Use child's name) *came to pick one and ate it for their tea.*

Activities using orange juice

- Using segments of satsuma or clementine, show how you can nibble a small hole in the corner of the skin and suck the juice out.
- Help your child to peel the segment open to see the tiny beads of orange inside. Have a go at squashing them on a plate to see the juice spray out. Now you have juicy fingers, paint some orange juice onto your lips.
- Slice a large orange into eight pieces and let your child squeeze their own juice into a cup. Praise them for the amazing juice they have made and offer them a straw or even a spoon to make it more interesting for tasting.
- If your child is not keen to drink the juice, ask them if they can stir the juice and kiss the spoon or catch a drip off the straw with their tongue. Remember to reward them with a high-five or praise for every small attempt.

HAVE FUN TASTING ORANGES

- Oranges are a great snack to have during bathtime because you can have lots of fun squeezing the juice out and not have to worry about the mess.
- Oranges can be quite acidic. Mix apple juice with orange for less acidity or try offering tinned mandarin slices with custard or yogurt on top.
- Freshly squeezed orange juice from the supermarket is great if you are unable to juice at home. Beware of pasteurized juice and added sugar.

Cabbage

In my opinion, cabbage should be renamed the superfood of the century. We should all be eating a serving of this fantastic vegetable at least once a week for its valuable cancer-fighting properties and vital sulphur compounds, something that our diets often lack, but which are hugely beneficial for our skin and joints.

Good for healthy skin and joints, cabbage also has anticarcinogenic (cancer-fighting) properties.

Cabbage is a powerful antibacterial and a great stress-buster.

Many of you may screw up your faces in disgust at the thought of something that stirs memories of smelly school dinners or your Grandma's overcooked Sunday roast. However, the reason your memory is so pungent

Folic acid
Vitamin C
Fibre

is because the cabbage you experienced was most likely soggy, overcooked and devoid of any nutritional value whatsoever.

As with many vegetables, the secret to great taste and gaining the optimum nutritional value from it lies in eating it raw or only lightly cooked. At Mange Tout there have been many surprised parents who have delighted at the fresh, nutty flavour of raw cabbage or been amazed at their child's suddenly acquired taste for grated purple cabbage.

Local farmers' markets, as well as your regular supermarket, will all stock a wide range of different cabbages: white, savoy, purple, pointed and miniature. Let your child choose two different ones to take home to look at, explore and compare.

Have you overlooked cabbage in the supermarket because it wasn't on your shopping list? Perhaps you've spotted it and not been inspired as how to cook or prepare it? The wonderful thing about cabbage is that it needs little or no preparation and your child can get involved in the process too.

A song to sing (to the tune of 'Row, Row, Row the Boat')

Roll the cabbage while sitting in a circle with friends.

Roll, roll, roll the cabbage
Roll it round and round
Cabbage, cabbage
Cabbage, cabbage
It grows in the ground.

Some games to play with your cabbage

- Take turns at peeling leaves off the cabbage. This can be made easier by cutting the stalk off the cabbage to release the leaves.
- Fan yourself with a large cabbage leaf or hide your face behind it and surprise each other.
- Can you wear a cabbage hat? This works particularly well with savoy cabbage leaves.
- What do the leaves feels like? What do they smell like?
- Does purple cabbage smell different to white or green cabbage?
- Wash the leaves in a bowl of water or you can drink water from a leaf that is shaped like a bowl. Use your cabbage bowl to put your sandwiches in at lunchtime or raisins in at snack time.
- Have a go at ripping, twisting and squeezing the cabbage leaves. Do they make a noise when you rub them together? Savoy cabbage makes a fantastic squeaky sound and feels just like dinosaur skin!

A rhyme to teach your child

(sort of to the tune of 'Old MacDonald Had a Farm')

Waving your cabbage leaves in the air:

> *Cabbages are good for me*
> *I'll crunch, crunch, crunch*
> *They're so yummy!*
> *Cabbages are good for me*
> *Some for lunch*
> *And some for tea!*

ACTIVITIES USING GRATED CABBAGE

Something that is very popular at Mange Tout is grated cabbage.

- Put some cabbage of your choice through the grating blade of a food processor and serve it in a cabbage-leaf bowl.
- Let your child sprinkle grated cabbage onto their plate and listen to the sound it makes when it drops.
- Bend your head down with your tongue sticking out so that it touches the grated cabbage on the plate and show how your tongue sticks to it like magic. If this is too daunting for your child, show them how a tiny piece can stick on its own too.

WAYS TO INCORPORATE CABBAGE INTO YOUR FAMILY'S DIET

- Cabbage loses most of its healing power through cooking. Therefore, a great idea is to simply flash cook it in a stir-fry.
- Cabbage is still best eaten raw. Add to a home-made coleslaw.
- Juice cabbage with carrot and apple for a tasty and nutritional thirst quencher.

Pineapple

Excellent source of vitamin C

Fibre

Rich in bromelain

When thinking of fruit to offer children, pineapple doesn't usually immediately come to mind. Perhaps this is because they can be a little costly and they definitely take a bit of time to prepare, but they are rich in nutrients and fibre and well worth taking the time to explore.

Pineapple is a delicious cure for digestive problems.

The enzyme bromelain is a good remedy for bruising because it breaks down the blood in the injured area that causes a bruise to appear.

The anti-inflammatory effects of pineapple are also great for sore throats.

The appearance of a whole pineapple very much resembles its tropical origins. The bumpy, prickly, tough skin and sharp, leathery leaves make us think of palm trees and golden sunshine. To discover the sweet, juicy fruit inside is like finding a treasure chest in the sand on a desert island.

OK, I know that's really over-the-top romanticizing, but you can make it feel that exciting and interesting for your child. For an adult who has experienced the circles of fruit soaked in sugar-laden syrup that can be served up easily just by opening a tin, the work involved in preparing a fresh pineapple may seem tedious. But to a child, it really can be a voyage of discovery.

BUYING AND PREPARING YOUR PINEAPPLE

- As usual, the voyage begins when you set out to buy the fruit. If your child has not experienced one before, describe a pineapple and see if they can spot one before you do. Let them handle it to feel the texture and the weight and smell it to check for ripeness. Unfortunately, this is not a fruit for rolling or playing with due to its nasty sharp leaves and rough skin, but the top, when sliced off, can be put on a warm windowsill in a saucer of water, where, if you are lucky, it will continue to grow.

- Most pineapples have a label attached with the country of origin. You could use this to explain how far the fruit has travelled before getting to you. Discuss how it may have travelled by boat or plane and talk about how long that may have taken. Explain how they grow (if you don't know, use the library or Internet to find out together). Respond honestly to your child's questions, enjoy the discoveries together and remember to share them later with other family members.

- The cutting and slicing of a pineapple is a job for an adult with a sharp knife, but this does not mean that your child cannot be involved. Ask for ideas of what colour it might be inside; whether it will be juicy or dry; will there be any pips?

- The core of a pineapple is tough and fibrous, but is edible if chewed well. Due to the high acid content, I would not recommend letting very young children chew on the core. My grandparents have told me that when they were in Jamaica, almost fifty years ago, workers in the pineapple canning factories used to lose their fingerprints or even the tips of their fingers due to constant contact with the fruit over long periods of time. I'm sure that things are better these days!

- Because the fruit is so juicy, it might be an idea to cut it on a tray so that the juice can be tipped into a cup for drinking. There are gadgets on the market to core and peel pineapples, but I find the best way is to slice off the top and bottom, halve it vertically and then cut each half into wedges. If the core is unpleasantly tough, then slice it off each wedge. Now slice off the fruit from the skin. Children will be able to cut the resulting slices into chunks.

Sing a song (to the tune of 'Twinkle, Twinkle Little Star')

Make a pineapple lollypop with a nice big chunk of pineapple, using a fork as the stick. Now you are ready to sing:

Pod says smell it
Pod says kiss it
Pod says lick it
Does it crunch?

WAYS TO EXPLORE AND ENJOY PINEAPPLES

- If this is the first experience with this fruit and your child is not adventurous about tasting new foods, then make this the end of the journey for them. Thank them for their help in preparing it and let them go to wash their sticky fingers (watch to see if they have a sly lick on the way).

- If your child is confident and happy to taste, then begin with the juice. This will have a concentrated flavour without complicating the experience with texture.

- While you cut the fruit, you could explain that pineapples are very fibrous and require chewing so that it will not come as a complete surprise.

- Pineapple is something that goes well with other foods and you can suggest experimenting with cheese and ham, using cocktail sticks for fun.

- Whiz fresh pineapple in a blender and freeze in ice-lolly moulds. It is better than ice cream for a sore throat.

- Small chunks make a colourful addition to fruit salad or fruit kebabs.

- Add chunks of pineapple to a home-made pizza.

Courgettes

At Mange Tout, and with many children when they are introduced to this member of the marrow family, a green courgette often gets mistaken for a cucumber. And quite rightly so because it does look rather similar. This similarity has in fact proved rather helpful during table time because children tend to be more forthcoming than usual in their willingness to try courgette. Although a slightly dry slice of courgette does not in the least bit resemble the watery, refreshing taste of a cucumber slice, it is rarely rejected. On one particular occasion, we had a one hundred per cent taste rate during the week that we introduced courgettes, spinach and bananas.

Courgettes have a high water content that means they are great for detoxing and diuretic action.

They are also a good laxative and a great anti-inflammatory.

The potassium and vitamin E help to counteract the negative effect of free radicals on the body.

Rich source of potassium
Vitamin C and E
Folic acid

Did you know? If you grow a marrow or courgette and score your name on the vegetable during its early stages of growth, the name will stretch and grow with the plant. See the growing section to find out how your child can grow their own marrows or courgettes.

Personally, I find courgettes that have been cooked or steamed rather watery and tasteless. Unless they are oven-roasted with lots of garlic and other interesting Mediterranean vegetables or perhaps scooped out and stuffed with a delicious bolognese sauce, they would not be top of my preference list. However, raw courgettes are a different experience altogether. Bear in mind that cooking removes quite a lot of the important nutrients in vegetables, so eating courgettes raw provides a powerhouse of goodness for you and your child.

Exploring your courgette

- Take time to look at and explore a courgette with your child. Notice how smooth and rubbery the skin feels and how prickly the stalk ends can be. Can you scratch the skin off and see what colour it is inside the courgette?
- Get your child to wash the courgette in some water or wipe it with a piece of damp kitchen towel so that you can do the following song and actions (make sure you have your own courgette too!)

 ## Sing (to the first line of 'Wind the Bobbin up')

Smell the green courgette (sniff)
Smell the green courgette (sniff)
Smell the green courgette. (big sniff)

Continue the song with the following words and actions:

Kiss the green courgette (kiss)
Lick the green courgette (lick)
Crunch the green courgette (make teeth marks)

How many different ways can you carry a courgette?

Have races to the end of the room and back carrying a courgette the following ways:

- Under your armpit.
- Under your chin.
- On your shoulder.
- In between your legs.
- Feeling brave? In your teeth!

Activity time

- Holding a slice of courgette, show your child how you can make teeth marks in the courgette or even bite it in half to reveal the shape of a smile or a moon. Encourage them to copy you or perhaps to kiss or lick the courgette slice.
- Using some sweet potato purée, tomato passata, hummus, soft cheese or avocado purée, make some courgette sandwiches with the slices by sticking them together with your chosen filling.
- Grated courgette is fun too. Make a face with courgette slices for eyes, a nose and a mouth, then use the grated courgette as stringy hair. This is a wonderful activity to introduce a child to the touch and texture of the vegetable if they are rather unsure about tasting at this stage. Getting them to enjoy a hands-on activity like this is a very positive experience for you both!
- Have a go at making a carrot and courgette caterpillar using slices of raw carrot and courgette threaded alternatively on to a couple of craft or match sticks. You could even use a cherry tomato for his head.

WHAT TO DO WITH COURGETTES

- The best way to take advantage of all the nutrients is to eat courgettes raw.
- Grate carrots and courgette and add sultanas to make a tasty finger food or to add to wraps or pitta bread.
- See page 163 for Pod's Green Soup made with courgettes. Make sure you have some raw courgette to dip in or some courgette sandwiches to accompany it. Surprisingly delicious!

Sarah Mitten, London

After trying doctors and nutritional experts, I was at the end of my tether last summer. My daughter who was eighteen months was making mealtimes a complete battleground. She would refuse to eat and most things would end up being thrown onto the floor. I was getting stressed out and needed help. Thankfully I saw a poster in the local butchers and as a last resort went along to the trial class last summer. I was overwhelmed by Lucy's enthusiasm and patience and signed up for the next term immediately.

We have done the class for a year and it has changed our lives. Amelia has more enthusiasm when it comes to food and her love of fruit is overwhelming. Amelia likes to help out in the kitchen and is often seen crunching on a carrot stick or mushroom while I'm making a Bolognese sauce. Even the supermarket can be fun. Amelia likes to put the fruit and veg into the trolley while munching on an apple or raw mushroom.

Lucy has made food fun and Amelia is often heard singing the songs. Lucy's ideas and, more importantly, encouragement for each child is fantastic. A huge thank you goes to Lucy and Pod for making our kitchen a calmer place.

Peppers

Looking at the beautiful, vibrant colours of the pepper, it is easy to imagine that the goodness is glowing from within them – and indeed it is. As with many fruit and vegetables, the goodness lies in their colour and you need to make sure that some of this colourful goodness gets into you and your child. There are many ways to enjoy peppers and a few children at Mange Tout have even been known to polish off a whole raw red pepper during circle time as if it were a delicious crunchy red apple.

The variety of rich glowing colours in which the many varieties of peppers come means that they are rich in protective antioxidants. These help us form the best defence against attack from internal and environmental bacteria.

Point out the colourful array of peppers in the supermarket and ask your child which is their favourite colour. Look out for miniature orange bell peppers, which are incredibly sweet. Choose one of each colour pepper to take home to look at and explore.

A rhyme about peppers

I like to eat peppers, crispy and sweet
Red, green or yellow, they're all good to eat
I'll eat them for lunch, I'll eat them for tea
I like to eat peppers, they're so good for me!

Activities using peppers

● Wash the peppers. Feel how smooth the skin of the pepper is by rubbing it on your cheek, then on your lips by giving the pepper a big kiss. Now try licking the smooth skin of the pepper. Remember to praise your child for all their engagement with the activity. Do all the peppers smell the same?

- Break open the peppers to reveal the secret seeds inside. Get your child to brush them out with their fingers onto a plate. Pepper seeds are not harmful if swallowed, however it is recommended that you avoid eating them because they have an extremely bitter taste and far more nutritional value lies in the flesh of the pepper itself.

- Feel how different the inside of the pepper is by stroking it with your finger or rubbing your tongue inside it. Scratch the inside of the pepper and see how much juice sprays out from it. Can you slurp the juice up? Fill a pepper half with some water and use it as a cup.

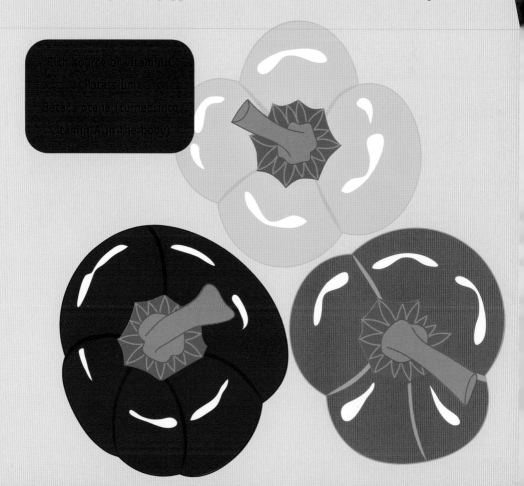

Rich source of vitamin C
Potassium
Betacarotene (turned into vitamin A in the body)

RECIPE IDEAS

- Peppers are an essential addition to any vegetable stir-fry.
- Offer colourful strips of sliced pepper with hummus or a soft cheese dip.
- Make a sweet pepper sauce to use for dipping things into or to have on some pasta. Choose red and yellow/orange peppers (these tend to be sweeter than the green ones), slice the peppers in half and deseed them. Rub with olive oil and bake them face down on a non-stick tray in a preheated 200°C oven for 30 minutes. Once they have softened and browned, the skin will come away from the flesh (this task can be a bit tedious, but little fingers love to pick so get your child busy peeling the skins off the peppers once they have cooled down). Purée the roasted peppers with a hand blender, adding some water for a runny consistency if desired.

Activity time

Use your fingers to feel if the sauce is hot or cold. Once the sauce is on your finger do some painting on a plate. Maybe give your fingers that are covered in sauce a really big kiss. Who can do the loudest kiss? Or suck your finger and pull it out with a loud POP sound. Encourage your child and praise them with high-fives and positive feedback.

Using the sauce and some raw pepper strips as paint brushes, have a go at painting your lips or even your tongue. Use a small hand mirror to help your child see what they are doing and to see the result.

Cauliflower

Vitamin C
Betacarotene
Folic acid

This is another vegetable that takes me back to my
school days and reminds me of the overcooked,
sloppy cauliflower cheese, which was sometimes my only lunch option
as a vegetarian. Like the Brussels sprout, this pretty white vegetable can
become bitter and very pungent when overcooked. Needless to say, in my
opinion it is far more enjoyable when eaten raw. Raw cauliflower has a very
delicate flavour with a slight nutty taste and a peppery hint. It makes a
delicious crunchy snack when dipped in some hummus!

If you have already introduced your child to cabbage, show them how
the leaves of a cauliflower can look a little similar, except that there is a
beautiful flower hidden inside. Have a look and see if you can find any
baby cauliflowers. Some supermarket chains stock them alongside other
baby and miniature vegetables. They are perfect for children to handle as
they are nowhere near as large as their ancestors and therefore not so
overwhelming. I also find them sweeter.

Cauliflower becomes unpalatable if overcooked
and loses much of its healing power.

The leaves should be added when
cooking for extra nutrients.

Studies have shown that lung, colon and
breast cancer are far less common in those
who eat large quantities of brassicas, such
as cabbage, cauliflower and broccoli.

Ways to explore and enjoy cauliflowers

- Have some fun pulling the leaves off a large cauliflower. You may need to score the leaves with a knife before attempting this as they can be very tough, but it is a great game to play, particularly with older, stronger children who enjoy the challenge. You could even turn it into a game of tug of war.
- Once you have uncovered the white, bumpy cauliflower, talk about how it looks like a fluffy flower. Stroke the cauliflower with your eyes closed. How does it feel?
- Once the leaves have been removed from the cauliflower, let your child have a go at separating the cauliflower into individual florets. Offer them a plastic knife to help break the thicker ones off. Using a knife yourself, slice through the centre of a piece of cauliflower and show your child how it looks just like a tree, with all its branches.
- Wash the pieces of cauliflower in a bowl of water. Show your child how the cauliflower is great for brushing teeth and it makes them really sparkle (the water will help with this magic). Listen carefully to how it squeaks when you run it over your teeth.
- Try crumbling some of the tiny cauliflower buds onto a plate and demonstrate, by licking them, how you can give yourself a spotty tongue.

Activity time

Explain to your child that cauliflower is extremely tickly. Tickle the palm of your hand with it and try it on your cheek too. See if you can tickle your tongue with a piece of raw cauliflower. Perhaps give it a really big lick because it looks just like an ice cream! Remember all the while that some zealous giggling will encourage your child to join in and play the game too.

A song to sing (to the first line of 'Wind the Bobbin up')

Pass the cauliflower, pass the cauliflower,
Pass the cauliflower.
Smell the cauliflower, smell the cauliflower,
Smell the cauliflower.
Kiss the cauliflower, kiss the cauliflower,
Kiss the cauliflower.

Another song to sing (to the tune of 'Twinkle, Twinkle Little Star')

Pod says smell it
Pod says kiss it
Pod says lick it
Does it crunch?

SOME COOKING IDEAS

- Steam or gently boil some cauliflower. If using baby ones, leave the small leaves intact and place it whole into a pan of boiling water for four minutes. If using a large cauliflower that has been cut into florets by your little helper, place some of the leaves into the water as they contain nutrients that can be absorbed by the cauliflower during cooking.

- Using the cooked cauliflower pieces, talk about the differences in colour, smell and texture compared to that of the raw florets. Practise using a knife and fork to slice and squash the cooked cauliflower. Suck the cooked cauliflower like a straw and see if any juice comes out.

- For a healthier cauliflower cheese, lightly steam the cauliflower then sprinkle with grated cheese and simply grill until crispy.

- Try breaking the cauliflower into tiny pieces and sprinkling on top of your soup. A favourite of mine is leek and potato soup with raw cauliflower croutons.

Apples

Apples are so much part of our everyday life that it's easy to forget that they may be just as new and scary to a child as any other fruit or vegetable. We are also so used to having them around and accustomed perhaps, to only buying our favourite type, that we lose sight of the breadth of choice available. As usual, the best place to introduce your child to a new fruit is in the supermarket or market. Point out all the different colours, shapes and sizes and talk about which ones may be sweeter, which ones crisper and which look as though the skin is easier to bite through. Try not to impose your knowledge or your own preferences, but you can say which ones you have never tried and suggest that your child selects a few to take home and try.

Once home, you can examine and discover each

Apples are a source of both soluble and insoluble fibre. Soluble fibre, such as pectin, helps to prevent cholesterol build-up in the lining of blood vessel walls, thus reducing heart disease.

The insoluble fibre in apples provides bulk in the intestinal tract, holding water to cleanse and move food quickly through the digestive system.

fruit in your own time. Explain that apples grow on trees and ask where the apple could have been attached to the tree. With slightly older children, you could talk about how the apple tree has blossom in the spring, which is pollinated by bees and insects, and from the blossom the little apples start to grow. Did you realize that the little bits at the opposite end to the stalk are the remnants of the blossom?

A rhyme to learn

Apples are great for rolling to and fro across the table or, even better, across the floor where they cannot fall and get bruised. Many of the rhymes for other fruit can be adapted for use with apples, but a very old favourite of my mum's is:

> *Do you like apples?*
> *Do you like pears?*
> *Do you like cuddling teddy bears?*
> *I DO like apples.*
> *I DO like pears.*
> *I DO like cuddling teddy bears.*

A game to play with family and/or friends

Before you try eating the fruit, make sure that your child has experienced it with all their other senses. Can you tell which is the green apple, for example, with your eyes closed, by feeling it, smelling it, even by licking it! Tongues are much more sensitive than fingers and a difference in texture may be more apparent this way. Use a blindfold and make this into an identification game that all the family can join in with.

Fun ways to explore apples

- Cut up the apples in different ways. Some apples, such as Granny Smiths and Red Delicious, are more difficult to cut than others (Golden Delicious and Cox's Orange Pippins tend to be softer). You will have to be the judge of how much your child can cope with, but once halved, the job of chopping them up becomes easier.

- A horizontal cut across the widest part of the apple will give a pretty star pattern in the centre (though cutting top to bottom makes coring easier). It's a good idea to leave the peel on, but if there are objections you could say something like, 'Why not peel it off yourself?' or 'Oh that's a shame because you washed it so beautifully.'

- On the subject of apple cores, children are very interested in why fruit have stones or pips and can happily accept that they are the way that trees have babies! But apple cores can be a reason why children dislike apples. The pip casings can get wedged between their teeth and be very uncomfortable – one boy I know used to call them toenails, urgh what an idea! So if your child is not happy to munch into a whole apple, cut one into quarters, remove the core and slice each quarter into three or four slices. Just offer a few at a time or share an apple between you (too much at once can be daunting).

Activity time

Because they are resilient, apples are an excellent fruit to let your child play shops with. You can balance them on scales, put them in paper bags or simply count, match and sort. Bobbing apples is a great bathtime game, helping your child to get water confident as well as food confident.

A counting rhyme to teach your child

Counting rhymes will familiarize your child with fruit as well as helping with their maths. You could try:

Five little apples, growing on the tree
I went and picked one and ate it for my tea.

Four little apples … etc. etc.

ENJOY TASTING APPLES

- Apple slices are great for dipping into purées of other fruit, such as raspberries or strawberries.
- Apples also go well with cheese and this makes a very nutritious and 'adult' snack for a youngster.
- Apples can be cooked whole in the microwave to make a different and very healthy dessert. Core an apple, leaving it whole, score around the middle with a sharp knife and microwave for a couple of minutes. The fruit will expand and the skin will probably part in the middle. Do not offer this to your child until it has had a good while to cool down. Variations can be made by putting sultanas, raisins or currants in the centre or perhaps other fresh fruit.
- Apples are such a handy snack to always have in your bag. Slice some and pop in a sandwich bag for a handy buggy snack.
- Stew apples with a teaspoon of brown sugar, a pinch of cinnamon and a couple of tablespoons of water. Blend and serve with yogurt or ice cream.

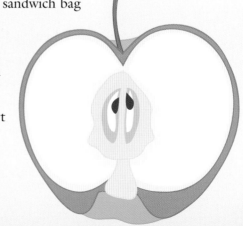

Strawberries

This must be one of the most popular fruits with adults, probably due to their natural sweetness and lovely bright colour. Years ago when they were only available in June and July, they were a seasonal treat and one to be savoured. English strawberries, freshly picked, are still unbeatable for taste and colour and the fruit that appears all year round in our supermarkets cannot compare.

I have fond memories of going to a pick-your-own farm on warm summer afternoons. Searching for the jewel-like fruits, hidden under leaves and protected from the ground by a layer of prickly straw, it was just like a treasure hunt. And it felt a bit sneaky – but so good – to sample the odd one as we picked. If you have the chance, this is by far the best way to introduce your child not just to this fruit, but to the whole idea of growing and harvesting.

If you cannot get to a pick-your-own fruit farm, then the summer is still the best time of the year to introduce strawberries to your child. They are plentiful in the supermarket, less expensive and you can have a lot of fun selecting where and how to buy them.

How to introduce your child to strawberries

- When you have got your fruit home, your child can help you give them a gentle wash in a sieve under cool running water. Explain that they need to be handled with care.

- Before eating or even preparing the strawberries, examine a few together.

- Note how the tiny seeds are on the outside, not hidden within like with apples, pears and many other fruit. Children may like to pick off a few seeds with their fingernails and see if they have any taste at all.

- Explain that as the seeds are so small and it would be a long and difficult job to take them all off, we don't bother.

- Smell the fruit, lick it with your tongue to feel the bumpiness of the seeds and roll it gently. Note that it is unlikely to roll off the table, as cherries or oranges might, due to its shape.

- Squeeze gently to discover what happens and lick your fingers.

- Show your child how to hold the fruit firmly, but gently. Gather up all the leaves and stalk in the fingers of the other hand and pull to remove the hull. This core of the fruit can be eaten too and you can have a discussion about whether it is better to leave them in or take them out.

The iron content in strawberries ensures the vitamin C is well absorbed into the body.

Strawberries are useful for both the prevention and treatment of anaemia and fatigue.

An excellent source of vitamin C.
Pectin (soluble fibre).
Iron.

Activity time

Strawberries are soft enough to be cut with a child's plastic knife and they look very different when cut horizontally from when they are cut across. If you stand a hulled strawberry on its widest part and cut downwards, you can create little heart-shaped slices. Use to decorate a plate or to create a pattern with other fruits.

A rhyme to learn together

You can learn this rhyme together and use it for lots of easy counting or even quite complex calculations.

Strawberries ripe and strawberries red
Strawberries growing in a warm straw bed
If I eat one and then pick three
There'll be plenty for my tea
But if I eat four and just pick two
There won't be enough for me and you.

A game to play

If you have plenty of small fruit, you could try playing this game (if you don't want to use fruit, then red buttons or counters will do).

Put two, three or more strawberries on the back of your hand. How many your child can manage will depend on age, size of hand and dexterity.

Say the rhyme:

Let's go searching in the strawberry patch
How many strawberries can you catch?

Throw the fruit up in the air and try to catch them as they come down. This is a variation on the old-fashioned jacks or fivestones game and can help with coordination as well as counting.

RECIPE TIME

These pretty little parcels of goodness shouldn't need anything sweet to help them down.

- Strawberries make wonderful smoothies, liquidized on their own or with other fruit.
- Strawberries are very satisfying to squash with a fork and then spread on bread or rice cakes. Delicious on top of soft cheese.
- For a healthy summer treat, liquidize strawberries with apple juice (no need to add sugar) and freeze in ice-lolly trays.
- Make strawberry 'ice cream' by mixing the mashed fruit with natural yogurt or fromage frais and then freezing.
- The flavour and sweetness of strawberries can be enhanced by a squeeze of lemon juice. Once your child is enjoying strawberries, you could experiment with your child to see if helps improve the flavour.

Bananas

A fantastic energy provider and a wonderful pit-stop snack that even comes in its very own disposable packaging. However, it doesn't matter how often we marvel at a wonder food like the basic banana, there are still a few of us who find them rather un-a-peeling!

For example, take my younger brother Owen who throughout weaning repeatedly refused bananas in any shape or form. Even at seven months, a jar of ready-prepared food would be refused if it contained banana. By the age of five, Owen was continuing to turn his nose up at anything even remotely banana-flavoured.

Once at a friend's house he was offered a banana and, not knowing what to do since he'd never eaten a whole one before, responded by saying, 'Oh, my mum takes all the seeds out first.' To which the

The easily digestible ripe banana contains soluble fibre, which is good for treatment of both constipation and diarrhoea.

The high potassium content helps prevent cramp.

Vitamin B6 is something often missing from children's diets and is known to help in the prevention of depression, skin problems and asthma.

Good source of potassium
Vitamin B6
Folic acid

mother replied, 'I don't think she does!' A banana is full of hundreds of tiny little seeds (almost like those from a vanilla pod). If you picture a slice of banana cake and think about all the tiny black dots speckled throughout, no one in their right mind (certainly not a mother with a hectic enough schedule) would take the time to remove every single seed. You'd be left with virtually no banana!

Bananas do have an extremely distinctive smell and texture and either of these factors can be a reason for your child's dislike of them.

Sing a song

(to the tune of 'My Ship Sailed from China' or say as a rhyme)

Stand nice and tall with your arms stretched above your head.

> *I am a banana, yellow and tall*
> *I'm really delicious*
> *But you can't eat me all*
> *I have a thick skin*
> *That's easy to peel* (bring your arms down as if you are peeling your
>   skin off)
> *I'm great as a snack* (clap, clap, clap)
> *Or part of a meal.*

Fun ways to discover/explore/touch and maybe taste

- Hide some bananas around the room and go hunting for them. Perhaps pretend to be hungry monkeys looking for food or elephants with long trunks searching out a tasty treat.
- Place a collection bowl in the centre of the room and once the bananas have all been found, have a go at making patterns on the floor by laying them down in a circle. Or make the shape of a train track or a ladder.

- Can you smell the banana through the skin?
- Use a knife to slice through the skin, cutting the banana into 'wheels' that you can peel.
- Use your finger to feel how slippery the smooth side of the banana is. Can you rub it on your lips or your tongue?
- Make sure that hands have been washed first, let your child have a go at peeling and mashing some ripe bananas in a bowl using their hands. It's extremely important that children have a chance to get messy and explore texture, particularly if you know it is something they will not yet put in their mouth. However, if you or your child would rather not use your hands, a fork or potato masher works well too. The mashed bananas will not go to waste as you can incorporate them into a tasty banana yogurt cake later – the mushier and browner the mixture the better (for the recipe, see page 173).

Activity time

Frozen banana is simply divine. Let your child prepare this and explain that they are going to help make a delicious ice-cream treat. All you need to do is let them peel and slice a banana (it doesn't matter how thin or thick the slices end up). Put them on a baking tray or some greaseproof paper and pop them into the freezer overnight. This is a fantastic evening activity as bananas contain a chemical that aids sleep. Also, if you wake the following morning to demands for the ice-cream treat, they are a perfect breakfast addition. They are also brilliant for popping into porridge and stirring through to help it cool down.

Leanne Brew, London

My son Samuel attended Mange Tout for a year from the age of two and a half. In that time he went from not even wanting to touch fruit and veg (unless they were mashed) to now eating a wide variety and being willing to try anything. The classes were a lot of fun with no stress about how much you tried, ate or joined in, but you were always rewarded and noted when you did. He loved going each week to see Lucy and Pod and talked at home about what he had done in the class.

As a parent, I learnt a lot about presenting fruit and vegetables in different ways, making it an enjoyable experience. It made me look outside my normal range of fruit and veg and I started to use a lot more variety in our diet.

Samuel is a great eater and I use a lot of the fun, games and praise learnt from the classes at home. It seems that if you get excited about fruit and vegetables, then children will naturally follow.

WAYS TO ENJOY BANANAS

- A favourite treat of mine is a banana and jam sandwich or banana, peanut butter and honey on corn crackers. Yummy!
- To make baked bananas, leave the bananas in their skins and wrap them in foil, getting your child to prick the skins with a fork first. If you want to make this an extra special treat, split the skin with a knife and add some small slices of good-quality chocolate. Bake in the oven or, even better, on top of a barbecue for 30 minutes. When the bananas have cooled off enough to open them, scoop out the mushy flesh and serve with yogurt or ice cream.
- Banana bread or cake is a great healthy snack.
- Children should only eat ripe bananas when the starch has turned to 'natural' sugars, otherwise they are not easily digestible.

Pomegranates

The word pomegranate means 'many seeded apple'. This is a rather strange and unusual fruit that looks a bit like a cross between a large pink apple and an onion, but the skin is much tougher and inedible. Look out for them in the shops throughout autumn time.

Pomegranates grow on shrubs and inside the fruit there are tiny little jewel-shaped pieces. Each bead encloses a nutty edible seed, which contains fibre and gives a crunchy contrast to the juicy sweetness of this fruit. Recently the pomegranate, and particularly its juice, has become very popular for its favourable healing and antioxidant properties. Whilst there is no doubt that one hundred per cent pure pomegranate juice is extremely beneficial to our health, it can also work out to be a rather costly addition to your weekly shopping budget. You can buy pomegranates individually for about the same price as a peach or a nectarine. Getting to grips with this wonderful fruit can be a real eye opener and great fun too.

A song to sing (to the tune of 'Frère Jacques'/'Are You Sleeping?')

Pomegranate, pomegranate
On the tree, on the tree
Ripe and sweet and juicy
Ripe and sweet and juicy
Just for me, just for me.

Ways to explore and enjoy pomegranates

- Feel the skin of the pomegranate and touch the spiky stalk end. Can you smell the pomegranate through the thick skin? Can you guess what colour the pomegranate will be inside?

- Slice the pomegranate into quarters and show how you can brush the tiny jewels out of the skin's shell using your fingers. Collect the little jewels in a bowl.
- Look closely at the inside of the leftover skin. Notice the pattern that has been left by the tightly packed little beads of fruit.
- Now look closely at an individual jewel of pomegranate. Can you see the magic seed inside? Squeeze the bead of fruit to release the juice and find the seed. Lick your fingers to see how sweet the juice is. Squeeze a jewel so that the juice sprays into your mouth. Does a pomegranate jewel disappear if we suck it like a sweet?
- Using a quarter slice of the pomegranate, squeeze it over a bowl and see how much juice comes dripping out. What colour is the juice? Dip your fingers into the juice and paint yourself some beautiful lipstick. If you suck the juice up a straw, does it change the colour of your tongue?

Sing a song (to the tune of 'Twinkle, Twinkle Little Star')

Pod says smell it
Pod says kiss it
Pod says lick it
Does it crunch?

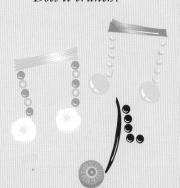

Excellent source of vitamin C
Excellent source of potassium
Polyphenols (antioxidants)

Pomegranates are perhaps the oldest fruit known to man and contain a good level of important nutrients, which have caused its recent resurgence in popularity.

The high antioxidant level helps eliminate harmful 'free radicals' from the body and enhances the immune function.

Juice from pomegranates may also help in the prevention of prostate cancer.

Activity time

Something that can be a little time consuming (depending on how nifty you are with a needle and thread), but that can make eating pomegranate really fun is threading the jewels onto some fine cotton to make a pomegranate necklace ready for sucking and munching. Try doing the same with blueberries or grape halves.

A song with actions (sing to the tune of 'I Had a Little Nut Tree')

Number rhymes and songs like this build numerical skills as well as introducing rhyme and rhythm to your child.

> *Five juicy pomegranates*
> *Growing on a bush*
> *Someone came and picked one*
> *And took it with a whoosh.*

Hold up five fingers and pretend to pick a fruit from each one in turn, putting the finger down when 'picked'.

Substitute the name of your child, mummy, daddy or a friend for 'someone'.

An activity to do together

- Draw a simple outline of a bush for your child to colour in green.
- Get them to put five pieces of fruit anywhere on the bush.
- Say the rhyme together and eat a piece of fruit each time (start with the adult taking the first piece with a big WHOOSH.)

Don't worry if your child doesn't want to eat the fruit. Keep the game relaxed and fun and count how many you each have had at the end.

WAYS TO INCORPORATE POMEGRANATE INTO YOUR FAMILY'S DIET

- The seeds can be eaten raw and are a good source of roughage to help cleanse the body. Try sprinkling on salads.
- To extract the juice, vigorously roll the fruit on a hard surface, breaking the juice sacs inside, before making a hole in the outer skin.
- Pomegranate makes a great citrus paste or marinade for use in cooking.
- Pomegranates form the base for the drink flavouring, grenadine.

Claudia Blandford, London

Christian was about nineteen months old when he first registered with Mange Tout. He was in the middle of some pretty intensive chemotherapy to treat his cancer and would not eat ANYTHING. While I am not trying to say that he miraculously started wolfing down Brussels sprouts and cabbage, he did start to express an interest in food, whereas before he had just been too scared of it. Nine months after starting Mange Tout (and finishing chemotherapy), Christian was a healthy, happy boy, eager to try most new foods placed in front of him. I say most, as he has no interest in baked beans or cabbage! I think this is a pretty amazing result seeing as he spent a lot of those nine months with a nasal gastric tube and me feeding him liquid vitamins through it.

Grapes

As a child, grapes were my absolute favourite fruit. It pained my mother to buy them though because when we were small, they were one of the most expensive fruits. No sooner were they hung decoratively over the apples and pears in the fruit bowl, they were instantly spied and demolished in a flash. Grapes (along with cherries) very early on became a special treat in our house.

Vitamin C
Excellent source of natural sugar
Antioxidant flavonoids

Grapes contain an enormous amount of protective compounds called polyphenols (concentrated in the skin of the grape – especially the black variety), which protect the heart and help prevent cancer.

Grapes are also useful in combating anaemia and fatigue.

However, as with all foods, not everything is to everyone's taste and you may well be baffled as to why your child is not tempted by a few sweet grapes, especially as they are such a great source of natural sugar and energy.

Activity time

My mum did manage to find a way to make grapes last just a little longer and it made us appreciate them even more. We used to slice the grapes in half and freeze them on some greaseproof paper. It was great fun sucking them like sweets, waiting for them to defrost in our mouths and then biting through the juicy fruit. Why not try it?

How to introduce your child to grapes

Make sure you have a mixture of red and green grapes at home. Try and buy seedless varieties because a child can easily be put off this delicious, sweet fruit if they crunch through a hard seed. However, if you can only buy seeded grapes, it is not a problem.

- Sit with your child and show them how to slice the grapes down the centre to reveal the seeds inside. Use your fingers to push the seeds out. How many seeds did you find? Kiss or lick the juice off your fingers. Feel how slippery the grape is inside. Does it squeak if you rub it on your teeth?
- Picking and washing grapes is lots of fun. Do they float or sink in the water?
- Show your child how to peel a grape. You may find it has them transfixed for ages!
- Try squeezing juice from the grapes onto a plate or into a bowl. You might want to cut the grapes in half first to make it easier and prevent the juice from spraying everywhere. What colour is the juice? Do green grapes make green juice and red grapes make red juice?
- Paint your lips with the juice or slurp it off a spoon. Explain how the wine that mummy and daddy drink is made from grapes.
- Explain to your child that raisins come from dried-out grapes. Try drying some out in a small container in the airing cupboard or boiler room (leave the container open for the moisture to escape).

> Always remember to supervise your child when eating grapes because they can easily choke on them if swallowed whole. It's always a good idea to slice them in half or quarters. Let your child have fun doing this with a plastic knife.

Quick reference and reminders

The following overview of the techniques used in this chapter can help jog your memory at a quick glance during or before a tricky mealtime. The ideas are simple enough to be incorporated into any situation that involves fruit and vegetables, be it play, food preparation or a mealtime.

If you are introducing a new food that is not outlined in the book, use the following examples to guide you and your child through the new experience. Start by choosing three examples that you think will appeal to your child, for example smelling, kissing and brushing teeth. Remember that with a completely new food, do try and do these activities away from the meal table and let your child work at their own pace.

Another way of using this section is to choose and prepare with your child any two or three fruit and vegetables and then follow this section as a game for all of them. This will help you to establish which techniques appeal to your child and encourage results.

Remember that exaggerated actions and enthusiastic examples will always encourage your child to join in. Any small amount of interaction or response from your child is positive progress and worthy of praise (high-fives work well in my classes) and certainly gives you something to build on in time. Forget that you want them to eat something and get them having some fun. Do not panic if your child does not follow suit with your first attempt, just continue the exercise and enjoy yourself.

CAN YOU SMELL IT?
What does it smell like? Do some big, deep breaths and exaggerated sounds, 'Mmmm, beautiful!'

CAN YOU KISS IT?

Can you do a baby kiss? Can you do a loud kiss? Can you do lots of kisses all over?

CAN YOU LICK IT?

Like an ice cream? Is it bumpy or smooth? Does it tickle your tongue? Can you paint your tongue and turn it a different colour? (You can paint your tongue red with beetroot!)

DOES IT CRUNCH?

Listen to see if it makes a loud or quiet sound? Is it hard or soft? Soft food makes a sound that only you can hear in your head. Can you hear it when you munch?

CAN YOU MAKE SOME SCARY OR FUNNY MONSTER TEETH?

Tuck green beans, carrot batons or pepper strips under your top lip to secure them in place. Once secure, put on your best monster noise along with googly eyes and encourage your child to join in. Can they scare your monster? Or make mummy laugh? When your child joins in, encourage them by saying how scary or funny they look. Demonstrate how your monster munches up his scary teeth, followed by:

> *'Where did my scary teeth go?'*
> *'Whoa, you're too scary, I'm frightened. Can you munch and crunch them all away.'*

Cover your eyes and explain that once the scary monster has gone, you can play another game. Don't worry if they do not eat the fruit or vegetables, at least the food had made contact with their mouth and tasting on a very basic level has taken place.

CAN YOU BRUSH YOUR TEETH?

'Open wide – ah, ah. Teeth together – cheese.'

Brush hard and ask if the food squeaks on your teeth? Can you hear a mouse?

CAN YOU SUCK IT HARD AND SEE IF THERE IS ANY JUICE INSIDE?

Leave a piece of fruit or vegetable hanging out so that it looks like a long tongue!

CAN YOU (OR YOUR MONSTER/RABBIT/LION) MAKE TEETH MARKS IN THE FOOD?

Demonstrate with vigorous munching and sound effects how you can see your own teeth marks in the food. Follow this with some enthusiastic exclamations to encourage your child to join in the fun.

CAN YOU GET IT TO STICK TO YOUR TONGUE?

This works best with raw and cooked leaves such as cabbage, spinach, lettuce or Brussels sprouts. Can you do some magic and get it to disappear? Demonstrate the magic yourself first.

CAN YOU MAKE A SMILE?

This works well with anything, but particularly slices of things such a carrot, cucumber, courgette, tomato, melon or kiwi. Take a bite to reveal the shape of a smile or even a moon.

SLICES OF FRUIT WITH THE SKIN STILL ON MAKE GREAT FUNNY LIPS

Slice them small enough so that your child can hold them in place with their teeth to make a funny mouth from the skin.

Always have a small hand mirror ready when your child is following some of the activities so that they can see to paint on some lipstick or to brush their teeth.

CAN YOU USE YOUR CARROT/MUSHROOM/BROCCOLI LIKE A BRUSH?

Put on some beautiful lipstick using your brush or make some monster lips. Sometimes it's useful if it's a food that's not been tried before to have some of it already puréed. For example, puréed asparagus with a lightly steamed asparagus stick to paint with?

FRUIT IS GOOD FOR SQUEEZING AND JUICING

Collect the juice in a bowl and use your fingers to paint lips with the juice or with loud noises just suck the juice off. Offer a straw to drink the fantastic juice you have just made.

HOW TO PREVENT MEALTIME MELTDOWN

There are numerous reasons why your child might not want to eat. Perhaps if they are ill or have been snacking too much throughout the day, then they may just not be hungry. However there are other factors to consider too and this section will help you to recognize them and offer some guidance in dealing with them.

The importance of language

Language, and the way in which we use it to introduce children to new or daunting experiences, is extremely important. This is true for most things, not only food. Following is an example from my own experience which shows how easy it is to forget that children pick up on the tone of our voice and our body language.

In my early days of nannying, I often found myself pushed for time at breakfast and, knowing I was already late for the school run, would try to hurry the kids eating up. However, no sooner did I begin the speeding up process with 'Come on, hurry up, we're going to be late,' than they seemed to take even longer with their meals. This situation used to drive me round the bend and my mornings would be a vicious circle of chasing my tail and never quite catching it. One day the penny dropped. The children sensed my agitation as I hurriedly threw a packed lunch together and raced to empty the dishwasher ready for the next dirty load. Then on hearing my voice rise further in a bid to hurry them up, their only response was to panic, stop eating and watch me dart around the kitchen like an uncontrollable tornado. A slight exaggeration, but you get the picture.

Situations like this can be made far easier for yourself with a little extra preparation. For example:

- Set the breakfast table and make a packed lunch the night before. Sit and eat breakfast with your child.

- Talk calmly about the plans for the day (even if this is only for your own benefit).
- Once you have finished, ask your child to help with some basic chores (packing away, wiping the table or helping to wash up or fill the dishwasher).

Always remember how important the tone of your voice is and the volume that's used when addressing children, especially if you want them to listen and follow instructions. Never forget that your body language speaks volumes too.

Children do not like surprises on their plates

Preparing your child for what to expect at the following mealtime is a good idea. On the way home from school or nursery, take some time to explain what you will be having for lunch or dinner. A great way to encourage a child to accept the meal that you have prepared is to offer them a simple choice so that they feel they are involved in the decision making. For example:

'When we get home we're going to have some yummy pasta and red sauce for lunch. What shall we have with it, peas or sweetcorn? Why don't you choose for us.'

When your child makes a choice, praise them for their decision and continue the conversation:

'Fantastic, what a great choice, I love sweetcorn. Perhaps we could have some cheese to sprinkle on too or maybe some grape juice to drink?'

If you are walking home, see if you can make a chant of the food you will be having:

'Sweetcorn, pasta, red sauce, YUM! Some for Lucy and some for MUM!'

Get your child involved

Once you get home, if you find that there is some time before the meal is ready, put out some paper and pens and ask your child to draw what you are both going to have for lunch. You could start them off by drawing round a dinner plate so they can fill in the food. Or perhaps if you have any white paper plates, let them draw their meal on the plate.

While you are preparing the food, talk to your child about their drawing. Ask them if they can draw the twisty pasta with a yellow pen or a big puddle of red sauce with grated cheese on top. How many pieces of corn can they draw? If you have time, sit down and draw a picture with them while the pasta is boiling.

Getting children involved with the preparation for a mealtime doesn't always have to include food preparation. You could ask someone to set the table and older children might like to make name settings. An activity that everyone can get involved in is table mat design using pictures of food from magazines or drawing pictures of favourite fruit and vegetables. The work can be laminated at a local printers or simply covered in sticky back plastic from an art shop. The mats will then be more durable and easy to keep clean.

Give plenty of warning

If your child is enjoying some free play or perhaps some television before a mealtime, make sure that you give them plenty of warning before they are expected to come to the table. It's a good idea to announce a ten-minute warning, followed by a five and finally two minutes in order to prepare them to leave what they are doing. I always find it useful to use the cooker timer for the final two minutes and explain that when the beeper sounds, it's time to come to the table (or go to the toilet, wash hands, then come to the table).

Sit down to eat with your child

Do make sure that you sit down to the meal as well. It is very easy to think of mealtimes as an opportunity to 'get a few things done' while your child is preoccupied with food. However, more often than not, your child is far more distracted by you emptying the dishwasher, answering the phone or hanging out the washing than interested in what's on their plate.

Make sure you have a plate with some of your child's food on it for yourself, even if it is only a handful of peas or some cucumber slices. If it's lunchtime, sit down while your child eats and prepare the vegetables for the evening meal. Talk about what you are doing, even if it's only peeling carrots. They may even surprise you and ask to try some. Even if you end up only crunching a few cucumber slices and telling a silly story, it creates a positive environment for your child to enjoy and feel at ease with food and comfortable with eating.

Also, do try to find time to eat together as a family (though it's not a good idea to enforce being together and then to argue over the meal). If there is only the opportunity to make it a once weekly occurrence, then grab that chance and the whole family will really benefit. As a teen, I used to resent having to come back from a friend's house on a weekend for Sunday lunch, but once I was at home enjoying my mum's cooking and catching up with everyone's news, it was a wonderful experience. I appreciate now why it was so important.

Understandably, sitting down at the table means that your attention can focus too much on what your child is or isn't eating. This in turn can raise your anxiety levels, which the child easily picks up on. The floodgates then open for an all-out battle of control between parent and child, resulting in frustration, uneaten food and a guilty conscience.

What else can you do?

Try turning out the lights or closing the curtains and lighting a candle in the centre of the table to create a calm atmosphere. Put on some classical music, rub lavender oil on your temples – whatever works to help you relax.

- Express your own preferences in a positive way. For example, 'These aren't my favourite, but I can eat them and daddy loves them.'
- Television and toys during a meal are not a good idea because they can be too much of a distraction from eating. Even if they seem like a great way of getting food into your child's mouth as they successfully polish off an entire meal without acknowledging what they are doing, television and toys remove the enjoyable social aspect of eating.
- I once looked after two children (three and five years old) who were not particularly interested in food. I would get them to tell me scary stories of spiders and witches while they were eating (of course I had to play along with the game and pretend to be very frightened), then rather casually I would interrupt and ask them to continue eating before they could carry on the story. On other occasions, we would talk about what we were going to do after dinner. An all-time favourite was discussing and deciding what to play during bathtime:

'Lets all have a munch of our pasta and talk about treasure hunts.' (Diving under the water to search for the pennies I threw in.)

'Let's all have a spoonful of yellow corn and talk about cake making in the bath.' (Lots of bubbles and mixtures in various cups and bowls.)

What about when mealtime meltdown happens?

If you are struggling and the mealtime meltdown has hit rock bottom, do not be discouraged.

- Grab a favourite story book and encourage your child to take a mouthful of food before you turn the page. Or perhaps, once they have eaten a mouthful, they can turn the page.

- If they are overtired and flailing, you could help them by creating the fork or spoonfuls, but do make sure they feed themselves. Do not fall into the trap of spoon feeding your child, especially when they are perfectly capable of doing it themselves. However, by all means help them along the way.

- Encourage them to feed themselves by pointing out interesting things in the book, even if it involves a rendition of:

  'Watch out! The hungry frog in the picture is going to grab your forkful of potato. Quick, hide it in your mouth so he can't get it.'

- Try telling an imaginative silly story. Make it up as you go along, pausing in places to calmly (in the same storytelling voice) encourage your child to have a munch of their sandwich or a slurp of their soup.

- Sometimes when children are over hungry and past feeling like eating, it's a good idea to talk to them about how their tummy needs some food to feel better or for some energy to play in the park later.

- Ask them what they like best on their plate and what colour it is.

- Get them thinking positively about the food on their plate:

  'Wow I think your tummy would love something orange. Carrots are orange like the sun, let's both have some carrot and see if it warms our tummies inside.'

- Sometimes just playing some classical music quietly in the background can help to create an air of calmness during a meal. The children I looked after would conduct the music with their arms in between mouthfuls – highly entertaining!

- A successful method my mum tried and tested on me during my terrible twos phase was that of simply leaving the room. If I refused to eat my food to try and stir a reaction from my mother, she would explain that she was going to empty the washing machine and leave the room. Unbeknown to me, she was watching me through a crack in the door as I eventually got bored of playing the 'I'm not eating game' and decided that there was only one thing for it – get on and eat. It soon became clear to my mum that I was simply trying my luck at asserting some control over her at the ripe old age of two!

Could tiredness be the problem?

Undoubtedly there will be times when a child is not trying to push your buttons, but is simply tired. Overtiredness can be an overriding factor in why a child is either not hungry or is generally disinterested in food. With our busy lives, hectic schedules and the uncompromising demands of modern life, mealtimes can sometimes get fragmented. Regular mealtimes for children are important so that they are ready to eat

Never leave a child entirely unattended with food in case of choking and if in a high chair, ensure that they are securely strapped in.

An early lesson

One experience in my early days of nannying that I will never forget was during my time with a family of five children. After a busy morning at the park during the summer holidays, all six of us were ravenous as we sat down to platefuls of vegetable lasagne and peas for lunch.

The eighteen-month-old twins in their high chairs had stinking colds and were not the slightest bit interested in their food. Rather than focusing on the other three who were all eating well, I stupidly threw a spanner in the works by naively offering the twins an alternative (I thought perhaps the cheese wasn't great for their colds and I wanted to make sure they ate something). This instantly sparked all-out war at the table, with the other three children all demanding something different and making excuses why they wanted toast instead of lasagne. Tears and tantrums followed and my blood pressure hit the roof. Twenty minutes later we were all tucking into lasagne with peas and toast! Although lasagne was always served with toast following this episode, we did manage to turn a corner and replace it with garlic bread, much to the kids' and twins' delight. Needless to say the scenario never repeated itself – I'd learnt my lesson.

when food is presented to them. Regular meals will also help to keep their metabolism at a steady rate.

If you are running late with a meal or sense that your child is getting hungry and restless, do offer them a small snack. Holding out until the meal is ready and expecting a tired, hungry child to get stuck in is not always the solution. Offering a few olives, some cucumber slices or a small plain cracker half an hour before their meal will not destroy their appetite. In fact, it can serve to increase their appetite or top up their energy levels so that they are ready and willing to sit down to a meal.

DOING MANGE TOUT WITH FAMILY AND FRIENDS

As well as doing Mange Tout at home with your child, how about getting some mums and toddlers together for a coffee morning and having a go at doing thirty minutes of Mange Tout as a group? Plan which fruit or vegetables you are going to do each week (two or three should be enough) and all bring your own produce or organize to share out the preparation tasks. Alongside the ideas from the individual fruit and vegetable sections, you can have a go at playing some of the following group games.

HIDE AND SEEK

Hide lots of produce to be collected in a basket or bag. Once collected, count how many of each item you have.

Hide one item and then use 'hot' or 'cold' to help the seeker (this works better for older children).

Hide a fruit or vegetable underneath a tablecloth. See if you can guess what it is just by feeling it.

BALANCING GAMES

Depending on the size, shape and weight of the fruit or vegetable, try balancing it on your head, shoulder, foot, chest or on a spoon or fish slice.

MEMORY GAMES

'I went to the market and I bought a ...'

The next person must remember what the first person bought and add their own item. The game continues as the list gets longer. Make sure you stick to buying fruit and vegetables.

Kim's game. Put a selection of fruit and vegetables on a tray and allow everyone one minute to look at everything. Cover the tray with a cloth or tea towel and see how many you can remember.

WORD AND SOUND GAMES

Instead of 'I Spy' use:

'For my treat I like to eat something beginning with...'

You can use all kinds of fruit and vegetables or foods and meals that contain them.

Can you think of a fruit that rhymes with the following:

- Bear
- Chapel
- Terry
- Reach

How many other words rhyme with the names of fruit and vegetables?

COUNTING GAMES

Use an old classic like 'Ten Green Bottles' and make up your own favourite song. For example:

Ten red apples, hanging on the tree
Ten red apples, hanging on the tree
Pick one apple and how many can you see?
There are nine red apples hanging on the tree.

Remember that things like peas in a pod, grapes on a vine, segments of an orange or seeds in an apple all make for interesting counting!

Other activities to do with friends and family

DECORATE A SWEET TREAT BOX

Use a plastic airtight container and decorate it with stickers and pictures to make a sweet treat box. Fill with a selection of dried fruits.

MAKE A STAR CHART

Make a simple chart or table to record the portions of fruit and vegetables consumed in a day. Get the whole family involved and see who can reach five a day or more. Perhaps you could issue gold stars.

DO A MEAL PLANNER

Prepare a meal planner for the week and help your child to draw the meals onto paper plates. Stick the planner somewhere that it can be clearly seen. Each day, ask your child to select the plate of food that they would like to have for their meal that evening and remove the plate from the option of choices for the following day. By the end of the week, not only will you have given your child a balanced diet, but they will have felt involved in the decision-making process too.

GO SHOPPING

On a shopping trip, get the children to help select the fresh produce – counting and colours can be used to help encourage them. If you can't tolerate the idea of taking the children round the supermarket, go to a farmers' market at the weekend or just a local fruit and veg market. Even a local greengrocer trip can be an adventure. Give a shopping list to each child, grouping the fruit and vegetables by colour or by beginning letter.

On a rainy day when the park is not an option, but you need to get out of the house, visit the supermarket without any intention of doing a shop.

- How many yellow things can you count?
- Find something big and something small.
- Ask your child to find a certain fruit or vegetable or something in a particular colour.

Let your child choose something new to buy and take home. Even though this is a reward for learning about so many fruit and vegetables, remember that they don't have to eat it. Instead plan how you might explore the new item. Perhaps by smelling, kissing or peeling it? Slicing it or squeezing the juice out? Cook and then mash it? Share it with mummy and daddy!

GET DRESSED

Getting children dressed can sometimes be a laborious task, particularly when children are unwilling to participate. Change their focus and ask them:

- 'What colour are your socks? White! Can you think of a vegetable that is white? Mushroom, yes well done!'
- 'What about your red t-shirt? What fruit is red? A strawberry, yes, shall we have some with our breakfast?'
- 'What about your blue trousers? I don't think any fruit or vegetables are blue are they? BLUEBERRIES! We can sprinkle some in our yogurt later. Daddy loves them and perhaps you can sprinkle his in too!'

FRUIT AND VEGETABLES ART AND CRAFT

If your child loves painting or colouring, let them draw or paint their favourite fruit and vegetables. You could also draw (or print off the Internet) some fruit and vegetable outlines for them to paint.

Once they are dry, cut the paintings up into simple squares or triangular shapes so that you can make a puzzle.

An older child might like to set up a still life on the table to copy. Let them arrange fruit in a bowl and decide where they would like to sit to copy it.

An older child might also like to make a papier mâché fruit bowl. Cover it with magazine clippings of their favourite produce and varnish it to seal.

Mother of Rhoslyn, 18 months

Mange Tout has helped my daughter Rhoslyn to associate food with enjoyment and fun rather than stress.

With Mange Tout, mealtimes are no longer a battlefield

It's a cliché, but Mange Tout revolutionized my daughter's approach to healthy eating. As a severe reflux baby she developed early negative associations with food. Later on, she would often not want anything to eat, let alone fruit and vegetables. Mange Tout's highly engaging and fun approach to the dreaded green things reversed her strong aversion to them and had her eagerly munching lettuce leaves and biting into juicy organic plums. She was not an easy convert, but Lucy's passionate belief in Mange Tout and her fabulous ways with food and stubborn young minds won her over.

Even now, a long time after she opened a mange tout pod for the first time, my daughter will always ask me to help her open one before she painstakingly picks out the little peas and pops them in her mouth.

I'm often told at mealtimes that a food, meal or way I'm doing something is just how Lucy or Pod at Mange Tout would do it. I'm so pleased the classes have had such an impact on her.

It's always amusing (well, maybe one out of three times) when children pull a face and resolutely refuse to try a new food. By encouraging children to not only smell and touch different foods, but also play games and have fun with them, Mange Tout demonstrates that they really aren't that scary.

My daughter was terrible with green vegetables, but I'll never forget a wonderful courgette 'broth' that she was served one day at Mange Tout. She actually asked for more. Not that strange a request from most children, but as the mother of a child with fairly bad eating 'issues', it was pure delight to see. It also demonstrated to me that there are countless ways to approach and enjoy vegetables, not just with a desire to throw them and your child out the window every mealtime!

Having pulled out nearly all my hair from the sheer frustration of my daughter's eating, or rather non-eating, Mange Tout was quite literally my crowning glory. It opened my eyes to the hundreds of ways in which you can make food, and in particular fruit and vegetables, not only a regular part of your child's daily diet, but something they actually want to eat rather than going 'yuck' at.

by Corrine McCaffery, London

GROWING
AND
GARDENING

One of the best ways to encourage your child to take an interest in fruit and vegetables and to enjoy the excitement of freshly picked produce is for them to grow their very own. Now, before you pass off the idea as being too difficult, rather time consuming or perhaps you think you don't have enough space – think again. If you have enough space for an empty eggshell, then you certainly have enough space to begin growing or sprouting something.

Helping your child to grow something and seeing them enjoy a sense of achievement through their efforts is hugely rewarding. Understanding the basics of plant science – how all living things need light and water to grow – will, without a doubt, help your child to appreciate the value and importance of fruit and vegetables. What's more, they may well surprise you and want to prepare and even eat the things they've grown themselves.

To grow your own fruit and vegetables, you do not have to be an experienced green-fingered gardener, nor do you have to turn the back garden into an allotment patch or convert your conservatory into a greenhouse. If you do not have a garden or much outdoor space, consider growing something small in a pot on a window ledge. Herbs do particularly well in or outdoors and only grow to a manageable size, depending on the size of the pot you use.

Before you start thinking about how, what and where to grow something, get your child excited and interested in plants and basic gardening. Begin with something that you may remember doing as a child at school: planting cress, sprouting beans or perhaps having a go at growing a sunflower.

Planting cress

Cress gives almost overnight results and reduces the impatience of waiting for something to grow, so it is a great activity to get your child started on. You can buy cress seeds at any home and garden store and even some supermarkets stock a range of seeds in the plant and flower section. Have a look and ask on your next shopping trip.

You will need to find a small container. It is usually best that it's not glass because you'll find that your child will want to handle the pot at regular intervals to check on the progress of the seeds. An empty yogurt or hummus pot or a plastic food tray is perfect. You will also need some cotton wool, toilet paper or kitchen roll to line the bottom of the pot.

- Get your child to pour enough water on the paper-lined pot or tray so that it all soaks up. This will keep the seeds moist and help them to sprout (germinate).
- Now, let your child sprinkle the seeds generously over the damp paper or cotton wool.
- Ask them if they think the seeds should be put somewhere dark or if they need sunshine or light to help them grow.
- Choose a warm light spot for the seeds and check them every twenty-four hours to see how they are doing.

Once the cress seeds have sprouted to a couple of centimetres or longer (help your child to measure the cress stalks with a ruler or tape measure) they are ready to be picked and eaten. Do not force your child to eat the cress, instead praise them for their great work and ask them if they would like to have a go at growing something else.

Grow your cress in an empty eggshell. The cress will grow like hair and you can draw a face on the eggshell to complete the character.

Sprouting beans

You can buy packets of dried beans very cheaply in the supermarket (butter and broad beans work particularly well). You will need a clear plastic bottle, some blotting paper (children's art or sugar paper works just as well) and some beans.

- Cut the top off the plastic bottle and cover any sharp bits with sticky tape.
- Roll the paper up and place inside the bottle, letting it unravel so that it expands to fit the bottle.
- Fill the bottle with water and then empty again so that the paper is nice and wet. Leave half a centimetre of water in the bottom of the bottle.
- Let your child slide three or four beans in between the bottle and the paper so that they can be viewed through the plastic. Make sure that the eye of the bean is facing upwards as this is where it will sprout.
- Push the beans quite close to the bottom of the bottle so that they can absorb the water. They will then be able to grow up the inside of the bottle and be viewed daily for any changes.
- Keep adding water once it has all been absorbed. They should start to sprout after a couple of days.

If you have great success with your beans and they outgrow the bottle, transport them to a flowerpot of compost and bury them five centimetres below the surface.

Show your child a potato or onion that has been left too long in the cupboard and that has started to sprout. Explain how potatoes grow from a seed potato and as it sprouts, it grows more and more potatoes. Digging for potatoes on my dad's allotment was like digging for buried treasure. Once you found one, you would find five or six.

You could also have a go at sprouting a mixture of dried beans, lentils and peas using the same method as with the cress seeds, except soaking the pulses overnight to soften them and speed up the sprouting process.

Growing your very own herbal medicine cabinet

Now that you have entered the world of simple growing, you can try your hand at helping your child to grow something a bit more adventurous, but explain that they will have to wait a little longer to see any results.

Herbs are a great addition to any kitchen and do not take up too much space provided you have a bit of room on your window ledge. If you are not a big fan of cooking with herbs, don't worry, they won't go to waste. In this section you will discover more reasons for growing them with your child, including a remedy for morning sickness.

YOU WILL NEED
- Some medium-sized flowerpots that have a drainage hole in the base.
- A small bag of compost.
- A plate, saucer or ice-cream container lid to stand the pots on so that the water can drain through and then be reabsorbed by the plant.
- A large spoon or fork (or perhaps treat yourself to a gardening trowel if you are getting into the spirit of things and plan to take your foray into the world of gardening further!)
- Some seeds of your choice.

WHICH HERBS?
Most supermarkets stock a good range of fresh herbs in small quantities. Select a few with your child to take home and explore. See how they feel, smell and taste using some of the techniques from the Quick reference and reminders section on pages 102–5.

Take your child to a local garden centre to have a look at all the plants and see if they have a herb section. You could show how to rub the leaves gently to release the aroma onto your fingers to smell. Decide which herbs you'd like to have a go at growing and see what range of seeds there is to buy.

A few herbs I would recommend that are easy to grow and great for cooking or alternative uses are:

BASIL
Fresh basil is not only delicious chopped into a tomato salad or stirred into hot pasta, but is also great for alleviating stomach cramps.

CHIVES
These are a great mild introduction to the rest of the onion family and are tasty when stirred into scrambled eggs or chopped onto new potatoes with a little butter.

MINT
An extremely versatile herb, which likes to grow in shade and not only aids digestion, but is also a refreshing pick-me-up when infused in hot (not boiling) water. A little girl I used to look after would always join me for a peppermint tea mid-afternoon if our energy was waning. Mint can also help relieve muscle tightness so throw a few leaves into your bath. A great morning sickness remedy that my friend swears by is mint, grated ginger and honey steeped in hot water for fifteen minutes.

ROSEMARY
Not only fantastic added to roast lamb and roasted sweet potatoes, it helps to stimulate circulation and has a reputation as a herb that can enhance memory. Next time you can't remember something, sit down and enjoy a cup of rosemary tea. I'm sure whatever it was will come flooding back.

THYME

Another easy herb to grow. Thyme enjoys a well-drained, sunny position and this herb's healing powers are intensified through drying. Hang a few sprigs upside down in a warm, dry place and once dry, infuse in hot water and drink as a tea to alleviate tension headaches.

LEMON BALM

A wonderful herb for children to grow because they will easily recognize the tangy, citrus aroma. It has potent antiviral properties, which can combat cold sores and used in tea it can help relaxation. Try using lemon balm oil on your temples to alleviate the blues.

 ## Planting the pots

- Ask your child to seek out some small pebbles or stones from the garden or at the park on your next visit. These will go into the bottom of the flowerpots to allow the water to drain, but prevent the compost from washing out through the hole.
- Let your child help to fill the pots with the compost, using a spoon or fork and their hands to press down firmly to compact it. Make sure you leave a four-centimetre gap at the top.
- Place the flowerpots on the tray and thoroughly water the compacted soil.
- Sprinkle in the seeds and then cover with another centimetre of soil.
- Place in a warm, light area and wait for the seeds to work their magic.

Growing your own fruit and vegetables

Now that you have mastered the art of pot planting, you are ready to take the final step to growing your own fruit and vegetables. Firstly, for ease and simplicity, see if you can buy a tomato plant. It will need regular watering and perhaps some liquid fertilizer. Your child will have great fun selecting the ripe tomatoes once they are ready for picking.

Other produce you could have a go at growing are peas, beans, strawberries, marrows, squashes and even radishes. All you will need are some empty cardboard egg boxes, compost, a growbag and some seeds.

- It might be a good idea to begin the germination process indoors. Fill each compartment of the egg boxes with compost and sow a seed two centimetres below the surface of each one. This is an easy task that you can get your child to do.
- Once the seedlings have pushed through the soil and have grown to about two to three centimetres in height, separate the sections of the egg boxes (this should be quite easy as the cardboard should be moist from the water. If not, cut and separate the egg box with scissors).
- Lay a growbag flat and use a fork to prick holes into the base of the sack. Children will have great fun doing this.
- Turn the sack back over and use a pair of scissors to cut a large oval or rectangular shape, five centimetres in from the edge of the sack, out of the plastic. The compost bag is now ready to be used as a miniature allotment.
- Use a spoon to dig holes into the compost big enough to fit the individual cardboard pots into, transporting the dug compost into a spare pot or bucket. Place the pots with their seedlings into the hole (the cardboard will eventually break down to a mush) so that they rest two centimetres below the soil level. Use the spare soil to cover them over.

Organic fruit and vegetables

For many parents, the question of whether organic food is better for their child is a concern. However, when it comes to eating fruit and vegetables, the important thing to remember is that it doesn't matter how or where the fruit and vegetables are grown because they are still highly nutritious. Some studies have shown evidence that there are benefits from eating organically grown fruit and vegetables, but this should not detract from the fact that foods grown by conventional methods are not unhealthy. They are hugely beneficial and contain all the nutrients that are found in organic foods.

Suzanne Marriot, London

My son Toby now treats his Mange Tout time as a second helping of lunch and will eat anything green or healthy. He will even suck on a slice of lemon if he's seen and heard about it from Pod. He can now recognize most fruit and vegetables in the market and shops and sees it as completely normal to have to wash and chop food to eat rather than expecting it out of a packet or the freezer.

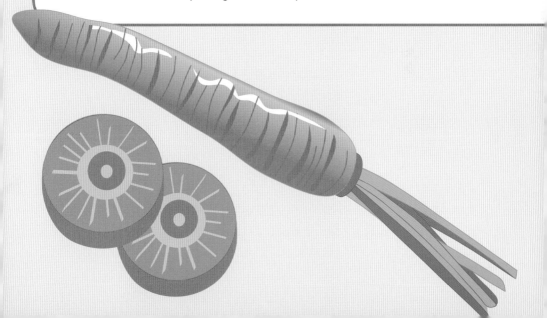

NUTRITIONAL INFORMATION

Fruit and vegetables are a fantastic source of vitamins and minerals. Encouraging your child to establish a diet rich in fruit and vegetables will guarantee a healthier lifestyle and reduce the risk of heart disease, cancer, stroke and many other diseases, both now and in your child's future. The increase in childhood diabetes is just one condition thought to be due to the poor diet of many children.

So what should my child be eating?

Young children live a very active lifestyle and the energy they require from food fuels not just this, but also aids rapid growth and brain development as well as the development of their immune system. Their dietary needs thus vary considerably to that of an adult.

The table below outlines the daily dietary needs of a child at different stages of life.

Age	Energy (calories)	Energy requirements from fat (%)	Protein (g)	Folate (µg)	Vitamin C (mg)
1–3	1,165–1,230	35	14.5	70	30
4–6	1,545–1,715	35	19.7	100	30
7–10	1,740–1,970	35	28.3	150	30

EAT LOTS OF VARIETY

It is always a worry for a parent to try and determine whether their child is eating enough, and eating enough of the right foods. The most important thing is that your child is offered a varied diet. A varied diet means a selection of different proteins, carbohydrates, fats, vitamins and minerals, which are available in a huge variety of different foods.

It is natural for children to dislike some foods, but provided they can gain their nutrients from others, there is no need to worry. For example, if you cannot persuade your child to eat broccoli, a great source of vitamin C and betacarotene, there is no need for concern because your child can get these nutrients from a range of other foods such as cabbage, leeks and parsnips.

Encouraging your child from a young age to experiment with lots of foods that have a variety of strong and diverse flavours and textures (not just fruit and vegetables, but herbs and spices too) will also make it far less likely that your child will turn into a fussy eater.

FIVE A DAY

All of us should be aiming for at least five fruit and vegetable portions in our daily diet as a bare minimum. In terms of portion size, a serving

Vitamin A (μg)	Vitamin B12 (μg)	Vitamin D (μg)	Calcium (mg)	Iron (mg)	Zinc (mg)
400	0.5	7	350	6.9	5
400	0.8	0	450	6.1	6.5
500	1	0	550	8.7	7

of fruit or vegetables is equivalent to what a child is able to hold in their hand.

If this seems hard to achieve, it is not a bad thing to disguise food your child doesn't like in different dishes. You can add all sorts of vegetables to spaghetti bolognese or encourage your child to eat certain fruits that they're not keen on by freezing them in small chunks to then enjoy as an iced treat. Bananas are fantastic sliced and frozen and taste just like banana ice cream. And if something like a soft-boiled egg doesn't tempt your child, you can try disguising eggs in pancakes or as eggy bread.

What about snacks?

As your child is growing and developing at an alarming rate, they will probably be constantly hungry, yet their small stomach is only able to tolerate small meals. When it comes to snacks, it can seem easier just to give your child a packet of crisps or a chocolate biscuit. There is no reason why they can't have these occasionally, but usually it is better for them to have something like a few crudités and a dip, a piece of fruit, a small sandwich or a yogurt. Snacks are a great way to keep your child's energy up if they are flagging and they will also improve their mood, but do remember snacks don't replace meals. And if your child is genuinely hungry, they will accept plain snacks.

The foundations for a nutritionally balanced diet

ENERGY

The best way to keep your child's energy up is through encouraging them to eat carbohydrates and fats. These foods are essential for your child's growth and development.

Carbohydrates include foods such as pasta, rice, bread, potatoes and cereals and these are all fantastic energy sources. These foods also contain other essential nutrients, for example an average-sized portion of potatoes contains half the daily allowance of vitamin C.

Small children may find the quantity of carbohydrates they need to ensure enough energy for the day a little bulky and this is why fats are another essential source

> For children, it is recommended that thirty-five per cent of their diet comes from fat, whereas for adults it is thirty per cent. It is however recommended that your child's energy intake from fat is no greater than thirty-five per cent because diets high in fats, in particular saturated fats (found in red meat, dairy products and processed foods such as crisps, cakes, biscuits and fast food) are not nutritionally beneficial.

of energy for children. Fats are also a great source of the fat-soluble vitamins A, D, E and K. As an adult, a low-fat diet can be beneficial, however for small children, a low-fat diet could mean they have an insufficient energy content.

There are many sources of naturally occurring fats so try not to give your child too much fat in the shape of crisps, chocolate and other junk foods. Meat and fish are good sources of saturated and unsaturated fats and also provide your child with essential nutrients including amino acids, iron, vitamins and minerals. Other great sources of fat include avocadoes, dairy products, eggs, nut and vegetable oils and nuts and seeds (although these should be avoided in children under five as they are easy to choke on).

Milk is a fantastic source of energy for your child as well as being a rich source of many essential nutrients. Breast milk contains more fat than cow's milk and so is a great energy-provider to start on. It is recommended that children over the age of twelve months drink whole cow's milk as their main drink. When your child is two, semi-skimmed milk can be introduced providing your child is getting enough energy from other sources. It is not recommended that your child drinks skimmed milk before the age of five as this will not provide enough energy and vitamin A.

FIBRE

Dietary fibre comes in two forms: soluble, for example oats, which can be broken down by the gut, and insoluble fibre, which cannot be digested by humans but is essential to maintain a healthy intestinal system. Children's small stomachs cannot cope with large quantities of high-fibre foods and it is thought that a child's diet that is high in fibre could lead to a reduced energy or nutrient intake.

Although children's fibre requirements are therefore lower than adults, it is still an important nutrient that can be found in wholemeal products, pulses, lentils, fruit and vegetables. Considering high-fibre foods are quite bulky for small children, particularly those with a small appetite, try giving your child a high-fibre snack when they're really hungry.

PROTEIN

Protein is also essential for your growing child and a diet low in protein can lead to poor weight gain, less height growth and learning delays. Meat and fish are particularly good sources of protein and providing your child with enough protein should not cause difficulties. It is also possible for children who are vegetarian and vegan to obtain enough protein in their diet, but more care must be taken. Dairy products and eggs are both good sources of protein, while pulses and beans also have a substantial protein content. Nuts, soya products and tofu are other good sources.

SUGAR

Refined sugar provides nothing but empty calories and is strongly linked to dental caries, which are prevalent in preschool children. It is important to limit your child's intake of sugary drinks, sweets and high-sugar snacks because a diet high in food that contains added sugars can encourage a sweet tooth.

Many fruit and vegetables contain naturally occurring sugars. These sugars are intrinsic, which means they are contained within the cell wall, whereas refined sugars are extrinsic (free), which means they are added to foods. We don't need to avoid intrinsic sugars in a healthy diet. Milk also contains sugars, but again these are not detrimental to health.

In addition to ensuring your child's diet is low in refined sugar, watch out for frequent snacking, which can again increase the risk of dental caries. The habit of daily and nightly teeth brushing should be encouraged.

SALT

It is particularly important that young children don't have a lot of added salt in their diet because their immature kidneys cannot tolerate large amounts. Salt occurs naturally at low levels in most foods, so it is not usually necessary to add more. Processed foods often contain high levels of added salt, so these should be avoided. If you are worried that the food you are serving your child doesn't have enough flavour, rather than adding salt try adding herbs and spices, which will also help broaden your child's diet.

If you do buy processed foods or meats such as sausages or bacon, always check the label. Salt is sometimes labelled as sodium because it is made up of sodium chloride. To work out how much salt is present, multiply the amount of sodium by two-and-a-half.

This table outlines how much salt your child can tolerate per day:

Age	Amount of salt (g)
7–12 months	1
1–3 years	2
4–6 years	3
7–10 years	5
Source: Food Standards Agency	

IRON

Iron is essential to the body as an oxygen transporter, taking it from the lungs to all the body's cells. Iron deficiencies are quite common in preschool children and can be associated with frequent infections, poor weight gain and developmental delay. The deficiency is usually down to the child's diet and may be due to inappropriate weaning foods or the early introduction of cow's milk, which doesn't contain iron.

Extended use of iron-fortified infant formula or follow-on milk may help, while lots of iron-rich food in which iron is easily absorbed should be included in your child's diet. Red meat and liver are great sources and can be consumed after the age of six months. These meats are not however always popular with children, so foods such as pâté and home-made hamburgers can be introduced from the age of one. Other great sources of iron include cereals, green leafy vegetables, pulses and white bread.

The absorption of iron is enhanced by vitamin C and food and drinks that are rich in vitamin C should be built in to mealtimes. Try fresh fruit juices, peppers, most fruits, Brussels sprouts or broccoli.

The trials and tribulations of feeding Charlie

When Charlie was a baby, he would breastfeed for only a few minutes before deciding he was full, tired or bored. Even on formula at five months, he never ever finished a bottle. But the lack of interest in milk didn't worry me, he was gaining weight (slowly) and so I reckoned it must be because he had a small appetite.

When moving onto solids the real fun began. He didn't seem to like anything. The only acceptable food was baby rice and pear (fine at six months, not so great at fourteen months) and yogurt. Everything else was refused. I attempted feeding one spoon sweet, the next spoon savoury when he was expecting sweet, but he sussed me out very quickly. I even tried mixing shepherd's pie or fish pie with yogurt to try and get it into him, but no joy. I used to blame teething, sick bugs, poor digestion, etc and he seemed to live off fresh air, not food. Charlie was walking at ten months, but still he was never hungry.

I decided to change tactics and make mealtimes fun times. Through play and complete distraction, he would eat! Success, but oh my, has my imagination been completely stretched over the years inventing new games for 'highchair playtime'.

Even now, aged three and a half, Charlie rarely says he's hungry and is still not interested in eating. He has amazing amounts of energy and to him, eating gets in the way of playing. I have learnt to live with his quirky eating habits and tiny appetite and I don't have to distract him to get him to eat anymore.

Mange Tout has been fantastic for us. There are many vegetables that Charlie doesn't like, but he wants to please Lucy so will do teeth marks, kisses or lipstick in those vegetables because he wants her praise and attention. He knows that he doesn't have to eat it and handling the food is enough. The Mange Tout sessions are a sophisticated extension of our highchair games. I never thought to play and explore with food in those early days and I wish Mange Tout had been around a few years ago. Lucy has managed to get an amazing array of vegetables close to Charlie's mouth and some he's even swallowed!

Having three children has proved to me that each child is an individual and no amount of parental pressure will change the quantity that a child needs to eat. But encouragement and fun can change the quality of what they eat and even the fussiest of eaters with the smallest appetite can be slowly introduced to new foods. Big thanks to Lucy!

by Caroline Frizzel, a Mange Tout mum

TREATS
AND BRIBES

'I wish my child never knew what a corn chip was.'
'My child will only eat yogurt and chocolate buttons.'

As an adult, we all understand the concept of a treat: a piece of cake on someone's birthday, a glass of wine and some salted nuts after a hard day in the office, a pizza out with friends or a bar of chocolate at the movies. There are also times when we overindulge, be it on holiday or maybe during times of stress. Nevertheless we do, as adults, understand the consequences of too much bad food.

Perhaps you find ways to counteract the effects of these damaging foods by incorporating exercise into your life. How many times do you have that doughnut on your coffee break and promise yourself an extra ten minutes in the gym? Or perhaps you balance out a Sunday morning fry-up by making a conscious decision to exceed your five portions of fruit and veg that day? Even I admit that there have been occasions when I've surrendered to the cookie jar only thirty minutes before dinner. Sometimes we just feel plain guilty for having devoured an entire tub of ice cream and promise ourselves we will eat better tomorrow.

But, and this is a big but, these are decisions we make as adults because we understand and are prepared to accept the consequences. I knew full well that eating a biscuit right before dinner was the wrong thing to do, however I could also judge for myself that it would not ruin my appetite for the vegetable stew I was impatiently waiting for.

Children, especially those under the age of five, cannot and do not understand why they are not allowed to eat treats all the time. A classic example of this was when my mother made the error of handing me a jam sandwich at lunchtime while I was having a complete meltdown and wouldn't eat anything. All mum wanted to do was ensure I had eaten something before I went to nursery so that I wouldn't be hungry. However, the jam sandwich incident came back to bite her for the three mealtimes that followed. I wouldn't eat my shepherd's pie, spaghetti bolognese or even cheese and Marmite sandwiches – I wanted strawberry jam!

Offering your child anything just because you want them to eat something can have a very negative effect. I'm not saying it's easy and I am the first to admit that my anxiety and stress levels go through the roof when the children in my care do not want to eat. However, opening the fridge and offering the child various alternatives in a panic for them to eat something will only reinforce in their mind that they are in control of their own food choices. All too soon this routine becomes familiar for the child and they begin to demand yogurts and biscuits and refuse everything else. It is really important that you explain to your child why certain foods are not very good for our bodies in order to encourage them to try foods that will keep them happy and healthy.

Bribes

'Eat your peas and then you can have some jelly.'

Initially, an innocent bribe can seem like a rather successful way to get your child polishing off the peas that have been left on the plate until last. However, bribing a child with the incentive of something sweet and delicious after they have eaten all their peas can create a wealth of problems and issues, especially in the way that your child views food as they grow and mature. Mealtimes for especially picky or selective eaters can turn into a battle of wills between parents and child.

However difficult a situation might be to work through and no matter how tired or exhausted you are feeling, it's never a good idea to bribe, punish or reward children to get them to eat. A bribe can actually cause the disliked food to become even more undesirable because the reward of sweets implies that the peas are 'bad'. As a result, the child will always expect a treat for eating them up and will probably always struggle to enjoy them.

CONSIDER THE FOLLOWING SCENARIO

Imagine you are attempting to discipline your child after asking them not to do something, for example jumping on the sofa, and you tell them that you now will not take them to the park. However an hour later, you realize that getting the child out of the house would be a blessing and a great way for them to burn off some excess energy, so you take them anyway.

The child has now learnt that even after doing something naughty, they still get to enjoy something fun. The position of authority that a parent holds is lost and the boundary of control is blurred. The likelihood of the child repeating the bad behaviour is also quite high because the discipline was not carried out. Therefore, if you want to discipline your child fairly and correctly, you must not warn them with empty threats that you are not able to follow through.

On the same note, if you want your child to accept and enjoy healthy foods, then you must not offer bribes or rewards for eating them because this only serves to increase negative attitudes towards the disliked food in question. It is also extremely important that you do not punish your child for not eating because this will again only create an even stronger negativity towards food and allow them to build associations of fear with food and mealtimes.

If your child doesn't want to eat what has been prepared as the family meal, then do not force them. However, avoid giving your child something else because this will only serve to reinforce the idea that every time they refuse a meal, they will be given something different. Your child will not starve after missing a single meal and if you stick to your decision, they will be

Studies have been carried out that prove that when children are given the opportunity to select and make their own food choices, they eventually, over a period of time, manage to balance out their own nutritional requirements.

ready to come and eat at the next mealtime. Offer fruit and vegetable sticks or plain crackers with water in the time between.

One way to help ensure your child is not hung up on junk food is for you to set an example by eating healthier home-cooked foods that are just as tasty and enjoyable. Make healthy food sound exciting. It is and it has far more colour and flavour than junk food. Good food is something to look forward to.

It is also important to remember that you can give your child the foods they want within a balanced and varied diet. Home-made burgers with potato wedges, thin-crust pizza with lots of veg and not much cheese and even a couple of pieces of good-quality organic chocolate are not bad.

Treats

If you would like to offer treats, make a conscious effort to implement some small changes so that they are healthy ones. This does not mean you have to substitute that iced doughnut with coloured sprinkles on for a dry wholemeal scone. Home-baked cakes and savouries are delicious and can be simple to make (see The Recipes chapter). You also have peace of mind that they do not contain any nasty chemicals or preservatives. Try designing your own wrappers for home-baked treats using kitchen foil and stickers or decorate an airtight container with pictures of their favourite television characters on to keep them in.

At the end of each Mange Tout class, we have sweet treat time from Pod's Magic Box. This is an airtight box decorated with colourful stars and a picture of Pod on the top. Inside there is a selection of different dried fruits. Have a go at making one at home and letting your child decorate it. There are many different varieties of freeze-dried and sun-dried fruits available in all supermarkets and good health-food shops, just be sure to check the labels and avoid those that have been produced with added sugar.

If you allow junk food, sweets and chocolate, make sure you place strict guidelines and limitations on them. Choose one day a week, perhaps one when you know that you will all be getting some exercise such as an afternoon bike ride or some swimming. Decide in advance what the treat will be and stick to it. When giving your child a treat, you could always break it down into smaller packets. That way your child is able to have a taste of what they like, but is also likely to accept that when their small bag is finished, it's 'all gone'.

I have fond memories of visiting a popular fast-food chain on the last Friday of every month. My two brothers and I would help our parents with the monthly shop, which involved two full trolleys and a seemingly never ending shopping list. But it was all worth it when we sat down together and enjoyed a much deserved and anticipated treat.

THE SUPERMARKET SHOPPING CHALLENGE

If you find shopping a challenge and want to avoid using bribes and treats, here are some tips for keeping your child's nagging under control:

- It's always a good idea to allocate jobs to older children who can fetch items off the shelves, tell you how much something costs (even add it all together on a small calculator) or tick the items off a list. Smaller children who sit in the trolley can be given a pad and pen for drawing (or they will always be trying to get their hands on your list).
- Make sure you visit customer services at the beginning of your shop and find out what number the confectionery and snack food aisles are so that you can avoid them, even if this means taking a longer route around the shop. If children are particularly persistent in their nagging requests or perhaps you need to visit the biscuit aisle to get some rice cakes, then keep their attention by talking to them or asking questions. For example:

    'How many things are in the trolley?'

'Who can spot someone wearing red shoes?'

'Can you hold a carrot under your chin? How long for?'

- If you have older children who don't even need a visual reminder on the shelf to spark their pleading, make them this offer. Explain that you are only going to buy the things on your list today, however when you get home your child can write their own list of all the food they would like for a birthday or Christmas party. Then, when that time comes, you will have a great idea of all their favourite things from which you can select a few. Each time you are on a shopping trip and your child makes yet another request for something you don't want to buy, remind them to put it on their special list when they get home.

Above all, if you really do not want your child to know what a corn chip, chocolate bar or a fizzy drink is before their fifth birthday, do them a favour and don't let them see you eating or drinking one. Better still, don't buy them. If you don't buy them, they can't have them and they certainly don't need them. Think of it as a great health kick for yourself and the whole family. You'll soon lose those few pounds that never seem to shift because those little snacks and treats you buy for the kids are no longer a temptation.

Kate Ash, London

A year ago, before starting Mange Tout, Lucy wouldn't eat any fruit or vegetables apart from bananas. Now, after two terms at Mange Tout, she eats avocado, apples, pears, kiwi and squash. She is still a fussy eater, but a million times better than she was. Coming to Mange Tout to see "Mange Tout Lucy" is one of the highlights of her week. She has also become a lot more interested in learning about fruit and vegetables and always wants to help me chop up veg for supper. Whenever I chop up the fruit, she wants to be the first one to see the pips inside.

THE RECIPES

These recipes (especially the soups and dips) have proved extremely popular when offered to children during our classes and at Mange Tout parties. When asking for the recipes, many parents have been taken aback by the sheer simplicity of them. The idea here is that parents will not be slaving over a stove for hours to then have their child throw the dish on the floor. Instead, these suggestions require minimal preparation, simple ingredients and are hassle-free for children to get involved with too.

Something as simple as stirring the contents of a mixing bowl can have a big impact on a child and help them to feel confident to sample something new that they helped to prepare and create. This is especially true if they have been praised for their efforts and encouraged by your enthusiasm. Similarly, helping to chop vegetables offers a perfect opportunity for touching, smelling or even crunching. Many fruit and vegetables, such as courgettes, mushrooms and tomatoes, are easy for children to help with and require only a plastic knife.

Many of these recipes can be offered as an accompaniment to a main meal and therefore the emphasis is on sampling a small amount of the new dish or eating it together with some more familiar food. For example, soup is extremely versatile and can always be served as an accompaniment to sandwiches, with pasta or used as a dipping sauce. Suitable for freezing, it will not go to waste and you can always modify it to your own taste (add a bit of parsley or pepper) and enjoy it yourself.

Sauces and dips

A simple salsa

Dips pair perfectly with tortilla chips. Try buying organic blue corn tortilla chips. Children will love the blue colour and blue corn is generally high in lysine, an essential amino acid, and zinc and iron. If you have blue and yellow chips, you can do a taste test and see which chip they like best.

serves 6–8

3–4 tomatoes, roughly chopped (this recipe tastes best when tomatoes are in season or if you use good tinned tomatoes)

1 onion, finely chopped

a bunch of fresh coriander, chopped

1 fresh red chilli, minced (optional)

1 tablespoon extra-virgin olive oil

juice of ½ lemon or lime

sea salt

● Mix all the ingredients together in a bowl and serve.

GET INVOLVED
For this recipe, your child can help mix the ingredients and choose some sprigs of coriander to garnish the salsa. Have them tear the coriander to see that herbs release flavour and scent when torn.

Guacamole

serves 6–8

4 medium-ripe tomatoes

4 ripe avocadoes, peeled and stoned

juice of 1 lime

½ small onion, finely chopped

2 garlic cloves, crushed

a bunch of fresh coriander, chopped

3 fresh red chillies, finely chopped (optional)

sea salt

- Use a knife to make a small cross at the base of the tomatoes and place in boiling water for approximately 1 minute. Remove from the water and peel the skin back from the cross. Remove the seeds and roughly chop.
- Scoop out the avocado flesh and process until smooth in a food processor or mash with a potato masher or fork. Stir in the lime juice.
- Add the tomato, onion and garlic to the mash. Stir in the coriander.
- If using the chillies, remove the stems and scrape out the seeds with a small sharp knife. Chop them finely and add to the avocado mash.
- Sprinkle in some salt and adjust the seasonings to taste. Cover tightly with clingfilm and chill for 1 hour before serving.

GET INVOLVED

For this recipe, your child might enjoy peeling and mashing the avocado. Also, you'll get more lime juice if you press down on the lime and roll it on the counter. Your child can help with this as well as using the juicer if you have one.

Hummus

This recipe is an old favourite and a great source of protein. Try serving with raw vegetable sticks or stuffed in hollowed-out cherry tomatoes.

serves 6–8

1 teaspoon cumin seeds

½ tin chickpeas, drained and rinsed

70g tahini (light works best), well stirred

50ml extra-virgin olive oil

1 teaspoon minced garlic

½ teaspoon sea salt

2½ tablespoons lemon juice

a pinch of cayenne pepper

- Briefly toast the cumin seeds in a dry frying pan until fragrant. Grind in a pestle and mortar.
- Add the ground cumin seeds with all the other ingredients to a food processor and pulse until smooth. If needed, add a little water to thin the hummus. Adjust the salt, lemon juice and cayenne to taste.

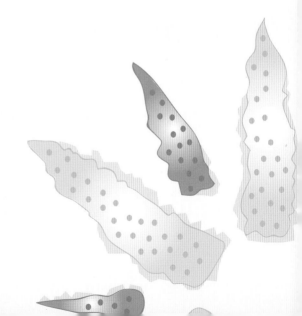

Veggie sauce

This veggie sauce is great with pasta, as a topping for fish or a dip for home-made potato wedges.

serves 4–6

3 tablespoons olive oil

1 large red onion, roughly chopped

3 large carrots, roughly chopped

2 large leeks, roughly chopped

3 red, yellow or orange peppers (or a mix), roughly chopped

2 courgettes, roughly chopped

1 garlic clove (not essential)

2 cartons of passata

sea salt and freshly ground black pepper

- Add the olive oil to a casserole dish and sauté the onion and carrot for 5 minutes.
- Add the leek, peppers and courgette and cook gently with a lid on until the vegetables are soft.
- Add the garlic, if using, and the passata and simmer with a lid on for 30 minutes. Take off the heat and season to taste.
- When the sauce is cool, place in a food processor and purée.
- Heat up and use straightaway or spoon the sauce into ice-cube trays or bags and freeze. Once frozen, empty the cubes into a freezer bag and use one or two as required.

GET INVOLVED

Get your child to rip open the peppers and take out the seeds. They can then just snap them into pieces for the pot because the contents are to be puréed at the end of the cooking anyway.

Salads and snacks

Crunchy corn salad

This raw salad packs a lot of nutritional punch and kids will love its crunchy texture. The recipe works best when sweet corn is in season. If it's not, try the recipe with tinned organic corn kernels.

serves 6

6 ears of raw sweet corn, kernels scraped off the cob

1 large red pepper, chopped

6 sprigs of fresh parsley, finely chopped

1–2 garlic cloves, minced

2 celery stalks, cut into small cubes

2 plum tomatoes, cut into small cubes

½ bunch of fresh coriander, finely chopped

1 tablespoon extra-virgin olive or walnut oil

1 teaspoon lime juice

sea salt

- Mix all the ingredients in a large bowl, adding salt to taste.

GET INVOLVED
Raw corn has a great crunchy flavour and releases a sweet juice when you bite into it. Children will love watching you scrape kernels off the cob, but be sure to hold the cob inside a bowl with high sides or the kernels will fly everywhere. Once each vegetable is chopped, let children scoop it off the cutting board and add it to the bowl. They can also help you measure out the oil and lime juice before adding them in and stirring.

Crispy lettuce rolls with almond dressing

These lettuce rolls make a great snack and kids will love assembling them.
If you don't have the exact ingredients listed below, substitute other raw
vegetables like cucumbers, tomatoes or corn or just use more of the few
ingredients you do have.

makes 12 rolls

for the lettuce rolls:

2 spring onions, finely chopped

1 avocado, peeled, stoned and cut into cubes

1 carrot, grated

1 celery stalk, finely chopped

½ red pepper, finely chopped

a handful of fresh sprouts (sunflower, alfalfa or bean)

12 large lettuce leaves

for the almond dressing:

4 tablespoons almond butter

4 tablespoons lemon juice

4 tablespoons water

sea salt and freshly ground black pepper

- Combine the first six ingredients in a bowl and mix well.
- To make the dressing, stir together the almond butter, lemon juice and
  water. Add salt and pepper to taste.
- Place a tablespoon of the mixture in the middle of a lettuce leaf and
  add a spoonful of the dressing.
- Fold the sides of the leaf into the middle, then fold the stalk end to the
  middle and fasten with a cocktail stick.

GET INVOLVED

Children will love stuffing and rolling the lettuce leaves. Just be sure they're
careful of the cocktail sticks.

Leek and courgette mini muffins

Children will love popping these bite-size muffins into their mouths. If you don't have mini muffin tins, a regular muffin tin will do. Enjoy them plain or serve with cream cheese, guacamole or another dip of your choice.

makes 36 mini muffins or 12 regular muffins

140ml olive oil

200g leeks, finely chopped

1 garlic clove, finely chopped

200g courgettes, grated and squeezed of excess liquid

300g self-raising flour

2 teaspoons mild curry powder

1 teaspoon paprika

250ml whole milk

2 medium eggs, lightly beaten

sea salt and freshly ground black pepper

- Preheat the oven to 190°C. Lightly grease two or three muffin tins (or cook in batches in one tin).
- Heat 2 tablespoons of the olive oil in a large frying pan. Sauté the leek until soft, then add the garlic, courgette and sauté for 2 minutes.
- Mix the flour, curry powder and paprika together in a large mixing bowl, then add the vegetable mixture and all the remaining ingredients. Mix together gently.
- Fill the holes of your muffin tin two-thirds of the way with the batter. Bake for 20 to 25 minutes until each muffin has a cake-like consistency.
- Take the muffins out of the tin and place on a cooling rack.

GET INVOLVED

This recipe presents plenty of opportunities for your child, measuring ingredients, squeezing excess liquid out of the courgettes, beating the eggs or filling the muffin tins.

Very berry couscous

As breakfast or a snack, this colourful fruit couscous is a great way to introduce yourself and your child to a new grain, not to mention the nutritional value of the berries and yogurt topping.

serves 4

185g instant couscous

250ml apple juice

250ml cranberry juice

1 cinnamon stick

2 teaspoons orange zest

750g mixed berries (raspberries, blueberries, strawberries)

200g Greek-style plain yogurt

2 tablespoons maple syrup or honey

- Place the couscous in a medium bowl.
- Pour the juices into a saucepan and add the cinnamon stick. Bring to the boil.
- Remove the juices from heat and pour evenly over the couscous. Cover the couscous with clingfilm and let it sit for approximately 5 minutes or until the liquid is absorbed. Remove the cinnamon stick.
- Separate the grains of couscous with a fork, then gently fold in the zest and the berries.
- Serve in bowls topped with a generous spoonful of the yogurt. Drizzle with the maple syrup or honey.

GET INVOLVED

After smelling the cinnamon stick, let your child drop it into the juice mixture. Involve them when you pour the juices over the couscous and ask them to observe how the couscous 'grows'. Children can help separate the grains of couscous and add the berries as well as top their own bowl with yogurt and maple syrup.

Light lunches

Pod's green soup

serves 2

3 large courgettes, thickly sliced

a large handful of spinach

½ sweet potato, peeled and cut into cubes

½ vegetable stock cube or 1 teaspoon Swiss vegetable bouillon powder

- Place all the vegetables in a pan with just enough water to cover.
- Add the stock cube or bouillon and bring to the boil.
- Simmer for 15 minutes, then blend in a food processor. Add more water until you get your desired consistency.

GET INVOLVED

Your child can slice the courgettes with a plastic knife.

Carrot and parsnip soup

serves 2

3 large carrots, peeled and chopped

1 large or 2 small parsnips, peeled and chopped (no need to take the cores out)

½ vegetable stock cube or 1 teaspoon Swiss vegetable bouillon powder

- Place the carrot and parsnip in a pan with just enough water to cover.
- Add the stock cube or bouillon and bring to the boil.
- Simmer for 10 minutes, then blend. Add more water until you get your desired consistency.

Leek and potato soup

serves 2

2 potatoes (preferably Maris Piper)

2 large leeks

1 small onion

25g butter

½ vegetable stock cube or 1 teaspoon Swiss vegetable bouillon powder

50ml milk (optional)

- Wash the potatoes and, leaving the skins on, chop into pieces the size of your thumb.
- Chop the tops and bottoms off the leeks and split down the middle to make it easier to wash out any mud and dirt. Wash, drain and roughly chop.
- Place the butter in a saucepan and melt. Add the leek with the onion and potato and cook, covered, for about 10 minutes until tender.
- Add the stock and enough water to just cover the vegetables. Bring to the boil.
- Simmer for 15 minutes, then blend in a food processor. While blending, add the milk, if using, or more water until you get your desired consistency.

Bubble and squeak patties

serves 2–4

250g cooked potato (with the skin on)

butter and milk, for mashing

sea salt and freshly ground black pepper

150g cooked cabbage, shredded (or you can use leftover cooked peas,
*   Brussels sprouts or broccoli)*

flour, for dusting

2–3 tablespoons olive oil

- Break up the cooked potatoes with a fork, add a little butter and milk and roughly mash. Season to taste.
- Add the greens to the potato and mix. Shape into patties and dust with flour.
- Heat the oil in a frying pan and fry the patties on both sides until golden brown.

Spinach soufflé

This recipe makes enough to give you and your child (six months and upwards) an enjoyable tea or light meal.

serves 2
125g spinach
4 tablespoons cottage cheese
2 eggs, separated
60g cheese, grated

● Preheat the oven to 190°C. Lightly butter a small casserole dish or two ramekins.

● Wash the spinach leaves and cook in just the water left on the leaves until wilted. Chop finely and mix with the cottage cheese.

● Beat the egg yolks lightly and stir them and the grated cheese into the spinach mix.

● Whisk the egg whites to soft peaks and fold carefully into the mixture.

● Pile into the prepared dishes and bake for 30 minutes.

Avocado pasta

Avocadoes are rich in healthy monounsaturated fats, which make them especially valuable for children. Weight-watching mums who think avocadoes are fattening are misguided. Half an avocado has the same number of calories as two apples and far more nutrients, so don't deprive your child or yourself on that basis. This pasta has a creamy, delicate texture that even young children will enjoy.

serves 4

2 ripe avocadoes (press the top of the pear shape to check they're soft and ripe)
juice of 1 lemon
1 fat garlic clove, crushed
sea salt and freshly ground black pepper
1 tablespoon olive oil
375g pasta

- Scoop out the flesh of the avocadoes into the bowl that you will serve the pasta from and mash to a cream. Add the lemon juice and mash again. Stir in the crushed garlic, add seasoning to taste and mash thoroughly to a creamy sauce.
- Cook the pasta in plenty of boiling water. Drain, pour on top of the creamy sauce, mix well and serve.

Bean chilli

The traditional beans for this dish are kidney beans, however you can
use any tinned variety here. To make this meal more fun, put bowls of
shredded lettuce, grated cheese and puréed avocado on the table with flour
tortillas or wholemeal pitta breads and get children to make their own
tortillas. It is messy but great fun! Alternatively, serve with rice, couscous,
pasta, mash or baked potato.

serves 2–4
1 tin kidney beans
1 tablespoon olive oil
2 large red onions, grated
1 carton passata
chilli powder, to taste

- Drain and rinse the beans in cold water.
- Heat the olive oil in a saucepan and sauté the onion for 2 minutes.
- Add the beans and cook over a medium heat for 5 minutes.
- Stir in the passata, cover with a lid and simmer for 25 minutes.

Supper sides

Carrot and almond jacket potatoes

Kids will love this slightly sweetened version of a baked potato.

serves 4

4 large baking potatoes

2 tablespoons olive oil, plus some extra to coat the potatoes

4 large carrots, shredded

75g almonds, chopped

1 teaspoon orange zest

1 teaspoon maple syrup

sea salt and freshly ground black pepper

butter

- Wash and dry the potatoes, then rub them in a little olive oil. Pierce them a few times with a fork and microwave for 5 to 10 minutes until cooked. If you have more time, you can bake them in a preheated oven at 200°C for about an hour and a quarter or until the flesh is soft.
- Place the oil in a frying pan and sauté the carrot gently for a few minutes.
- Add the almonds and sauté for another 2 minutes.
- Add the orange zest, maple syrup and salt and pepper to taste. Continue to cook until the carrots are soft.
- When the potatoes are cooked, cut a deep cross in the top of each one.
- Open the potatoes, mash in the butter and sprinkle a little sea salt onto the flesh before adding some of the carrot mixture.

GET INVOLVED

Let your child wash, dry and rub the potatoes in olive oil. If they are older, let them pierce the potatoes. Allow them to cut open and top their own potatoes.

Honey-roasted root vegetables

This recipe makes great use of the fantastic selection of autumn and winter root vegetables.

serves 8

1 large sweet potato, peeled and coarsely chopped

2 turnips, peeled and coarsely chopped

2 parsnips, peeled and coarsely chopped

3 shallots, halved

2 tablespoons olive oil

175g honey

½ teaspoon sea salt

- Preheat the oven to 225°C.
- Combine all the ingredients in a large bowl and toss to coat the vegetables evenly with the oil, honey and salt.
- Place the vegetables in a roasting tray. Bake for 35 minutes, stirring every 15 minutes, until the vegetables are tender and beginning to brown.

GET INVOLVED

Encourage your child to examine the similarities and differences in the vegetables – their colour, texture and how they grow, etc. Allow them to taste the honey before they toss the vegetables in it.

Desserts

The whole family will enjoy the following desserts and you'll enjoy the healthy inclusion of fruit and vegetables.

Courgette walnut cake

This cake is so tasty that children won't even realize there's a vegetable in their cake and it gives you a unique way to use up all those extra courgettes from the summer garden.

serves 8

a large handful of walnuts

220g plain or wholemeal flour

2 teaspoons ground cinnamon

1 teaspoon sea salt

1 teaspoon bicarbonate of soda

1 teaspoon allspice or ground cinnamon and nutmeg

½ teaspoon baking powder

340g sugar

300ml vegetable oil

3 eggs

1 tablespoon vanilla extract

zest of 1 lemon

2 courgettes, coarsely grated

- Preheat your oven to 165°C. Grease a medium-sized cake tin with butter.
- Spread the walnuts on a baking tray and toast in the oven. When cool, coarsely chop.

- Whisk the flour, cinnamon, salt, bicarbonate of soda, allspice and baking powder in a bowl to blend.
- Whisk the sugar, vegetable oil, eggs, vanilla and lemon zest in another large bowl to blend. Whisk in the flour mixture. Stir in the courgette and walnuts.
- Pour the batter into the prepared tin. Bake the cake for about 1 hour and 15 minutes until a skewer inserted into the centre comes out clean.
- Let the cake stand for 10 minutes, then turn out onto a rack and cool completely.

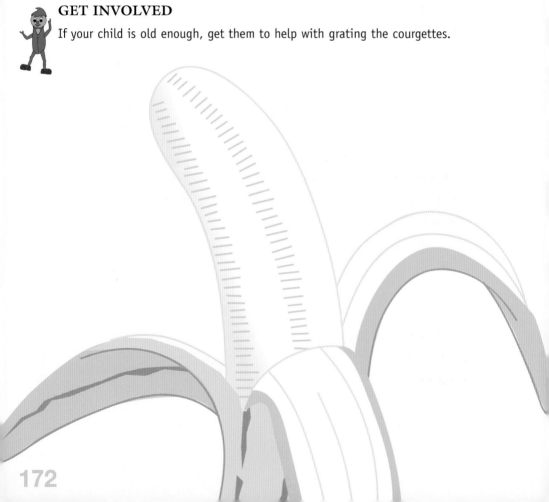

GET INVOLVED

If your child is old enough, get them to help with grating the courgettes.

Banana yogurt cake

Always a favourite, this banana cake has an especially moist texture thanks to the addition of calcium-rich yogurt.

serves 10
120ml sunflower oil
225g light muscovado sugar
3 eggs
225g self-raising flour
1 teaspoon ground cinnamon
2 ripe bananas, peeled and sliced
150ml natural yogurt

- Preheat the oven to 180°C. Grease a loaf tin with a little oil and line the bottom with baking paper.
- Pour the sunflower oil into a large bowl. Add the sugar and whisk for a few minutes. Add one egg and whisk until the egg is fully integrated. Repeat with the remaining two eggs.
- Gently fold the flour and cinnamon into the mixture.
- Mash the bananas and then fold them into the cake mixture along with the yogurt. Pour the mixture into the tin.
- Bake for 1 hour, or until the cake is risen and golden.
- Let the cake stand for 10 minutes, then turn out onto a rack and cool completely.

GET INVOLVED
Children will love mashing the bananas and cracking the eggs (just be sure to check for shells).

No-bake peaches-and-cream cups

Not only does this recipe require no baking, it's also dairy and wheat-free, making it perfect for children with allergies. If your child doesn't like peach skin, they can be peeled like tomatoes.

serves 6
for the crust:
220g almonds
180g pitted dates
for the cashew cream:
150g cashew nuts
50ml maple syrup
seeds scraped from 1 vanilla pod
1 teaspoon grated fresh ginger (optional)
70ml water
juice of ½ lemon
for the filling:
2 large peaches, stoned and sliced

- In a food processor, process the almonds into a fine meal. Add the dates and process again until you have the consistency of sticky dough.
- Line the bottom of six ramekins with a thin layer (about 3mm) of the dough. Freeze for 30 minutes.
- While the ramekins are in the freezer, make the cashew cream by puréeing all of the ingredients together in a blender.
- Spoon the cream into the ramekins and top with the peach slices. Eat!

GET INVOLVED
Under adult supervision, children will enjoy pushing the buttons of the processor and using their hands to press down the mixture into the ramekins.

Treats

For parties or special occasions, melt some good-quality chocolate (seventy per cent cocoa) and get the children to dip and coat their favourite fruits in it. Set them on greaseproof paper in the fridge. Strawberries and orange segments (with the fine skin removed) work well.

Other great ideas for healthy treats include:

- Home-made popcorn.
- Corn thins with avocado.
- Fruit kebabs (children can have a go at making their own).
- Sugar-free jelly set with fresh fruit.
- Home-made wraps using large lettuce leaves with grated carrot, courgette and hummus inside.
- Dried fruit such as pineapple, mango, apricots, pears or papaya.
- Yogurt-covered raisins, apricots and bananas (available from all good health-food shops).

Acknowledgements

Jacq, you are the most practical, pro-active, kind and patient woman I have ever worked with. I can't thank you enough for all the hard work and support you've given me. Thank you for confirming my belief in Mange Tout and for the positive influence you have had on our success.

I am extremely grateful to Claudia Blandford for demonstrating that 'word of mouth' could be an invaluable marketing tool for Mange Tout!

Mum and Dad, thank you for being the most encouraging, supportive and committed parents ever. Dad, your patience and perseverance never ceases to amaze me and Mum, where would I be without your creative flair and inspiration? I love you both very much.

Pete, we make an awesome team. Without you there would be no Mange Tout. You are my Rock, and you rock my world!

Helen, for whom no mountain is too high, no task too tricky and whose enthusiasm knows no bounds – you're a star!

Cath, for making our Oxford branch a great success and being an invaluable point of reference.

Rich, thank you for putting up with my kitchen clutter, morning clatter and computer conundrums.

Paul, for having humungous patience to not throttle me the zillionth time I talked shop! Your expertise in retail therapy is also much appreciated!

Melinda, I award you a medal for 'morning sickness survival' in honour of you helping to create such culinary delights, even when food was the furthest thing from your mind.

I also hold the utmost appreciation to all our Mange Tout members for supporting us right from the very beginning. Thank you all.